OIL PRICES

FUNDAMENTALS, FLUCTUATIONS AND IMPACTS

ENERGY POLICIES, POLITICS AND PRICES

OIL PRICES

FUNDAMENTALS, FLUCTUATIONS AND IMPACTS

JASPER C. MULLEN
AND
BLAINE M. LYNN

EDITORS

Nova Science Publishers, Inc.
New York

For permission to use material from this book please contact us:
Telephone 631-231-7269; Fax 631-231-8175
Web Site: http://www.novapublishers.com

NOTICE TO THE READER

The Publisher has taken reasonable care in the preparation of this book, but makes no expressed or implied warranty of any kind and assumes no responsibility for any errors or omissions. No liability is assumed for incidental or consequential damages in connection with or arising out of information contained in this book. The Publisher shall not be liable for any special, consequential, or exemplary damages resulting, in whole or in part, from the readers' use of, or reliance upon, this material. Any parts of this book based on government reports are so indicated and copyright is claimed for those parts to the extent applicable to compilations of such works.

Independent verification should be sought for any data, advice or recommendations contained in this book. In addition, no responsibility is assumed by the publisher for any injury and/or damage to persons or property arising from any methods, products, instructions, ideas or otherwise contained in this publication.

This publication is designed to provide accurate and authoritative information with regard to the subject matter covered herein. It is sold with the clear understanding that the Publisher is not engaged in rendering legal or any other professional services. If legal or any other expert assistance is required, the services of a competent person should be sought. FROM A DECLARATION OF PARTICIPANTS JOINTLY ADOPTED BY A COMMITTEE OF THE AMERICAN BAR ASSOCIATION AND A COMMITTEE OF PUBLISHERS.

Additional color graphics may be available in the e-book version of this book.

Library of Congress Cataloging-in-Publication Data

ISBN: 978-1-61942-485-2

Published by Nova Science Publishers, Inc. ✛ *New York*

CONTENTS

PREFACE

High oil prices affect nearly every household and business in the United States. During the course of 2008, oil prices doubled to more than $145 per barrel and then fell by 80%. In early 2011, there was a run-up of about 20%, sending gasoline prices to near 2008 highs. This book provides background on financial speculation in oil, the workings of oil derivatives markets, and the different types of firms that trade in those markets. Also discussed are the concepts of manipulation and excessive speculation, and the fundamental factors that affect oil prices.

Chapter 1 - High oil prices affect nearly every household and business in the United States. During the course of 2008, oil prices doubled to more than $145 per barrel and then fell by 80%. In early 2011, there was a run-up of about 20%, sending gasoline prices to near 2008 highs. Few would rule out the possibility of similar price swings in the months to come. What explains oil price volatility?

Some consider price movements such as those of 2008 and early 2011 to be more extreme than warranted by the fundamentals of supply and demand. Their explanation for unstable commodity prices focuses on financial markets for derivatives contracts linked to the price of oil—futures, options, and swaps. Many market participants are pure financial speculators, who never deal in physical oil, but earn large profits if they can correctly forecast price trends. Critics claim that such traders can drive oil prices above fundamental levels, resulting in a "speculative premium" that imposes unjustified costs on consumers. Although the relationship between speculation and commodity prices has been studied extensively, consensus has not emerged as to whether speculative trading causes unusual oil price volatility.

Chapter 2 - Oil prices generally increased from 2002 until mid-2008, collapsing with the economic downturn and then rebounding with global economic recovery in 2009 and 2010. In 2011 to date, the price of West Texas Intermediate crude oil has ranged from less than $80 to more than $110. These price levels and fluctuations stand in contrast to the relatively low and stable oil prices of the 1990s, when the level of WTI averaged about $20 per barrel. Periods of price increases raise economic and energy security concerns. This report provides an overview of factors that contributed to these oil price movements and may continue to drive prices in the future.

Chapter 3 - Petroleum prices have risen sharply since September 2010, at times reaching more than $112 per barrel of crude oil. Although this is still below the $140 per barrel price reached in 2008, the rising cost of energy was dampening the rate of growth in the economy during the first half of 2011. While the price of oil has increased sharply, the volume of oil imports, or the amount of oil imported, has decreased slightly. Overall resistance by market demand to changes in oil prices reflect the unique nature of the demand for oil and an increase in economic activity that has occurred since the worst part of the economic recession in 2009. Turmoil in the Middle East was an important factor causing petroleum prices to rise sharply in the first four months of 2011, which could add as much as $100 billion to the total U.S. trade deficit in 2011. The increase in energy import prices is pushing up the price of energy to consumers and could spur some elements of the public to pressure the 112th Congress to provide relief to households that are struggling to meet their current expenses. With oil prices rising to over $100 per barrel in early 2011, the International Energy Agency cautioned that the rising price of oil was becoming a threat to the global economic recovery. This report provides an estimate of the initial impact of the changing oil prices on the nation's merchandise trade deficit.

Chapter 4 - Rapid changes in the price of oil and the impact of such price changes on economies around the globe have attracted considerable attention. In mid-2008 as the price of oil rose to unprecedented heights and then dropped sharply, the international exchange value of the dollar fell and then rose relative to a broad basket of currencies. For some, these two events seem to indicate a cause and effect relationship between changes in the price of oil and changes in the value of the dollar. Despite common perceptions that there is a direct cause and effect relationship between changes in the international exchange value of the dollar and the price of oil, an analysis of data during recent periods indicates that changes in the price of oil are driven by changes in the demand for oil that is different from the supply of oil, rather than changes in the value of the dollar. The rapid increase in oil prices in early 2011 reflects rising demand for oil and other commodities and uncertainty in global markets keyed to political turmoil in North Africa and the Middle East.

Chapter 5 - With gas prices now at more than $4 per gallon, Rep. Elijah E. Cummings, the Ranking Member of the House Committee on Oversight and Government Reform, asked minority staff to examine the fundamental causes of recent price increases. Staff reviewed the work of ten different congressional committees, including the Oversight Committee, and analyzed data and information from a number of experts, including industry representatives, government officials, and academics. This report presents the results of this review.

The report's chief conclusion is that, in order to make the most significant impact on lowering gas prices, the Committee's primary focus should be on countering the growing impact of excessive speculation, rather than pursuing the oil industry's priorities of increasing domestic drilling or repealing safety measures put in place after the devastating BP oil spill. Experts estimate that excessive oil speculation could be inflating prices by up to 30%, while increasing domestic drilling would impact prices by only about 1%, and then only after a decade or more. Addressing excessive speculation offers the single most significant opportunity to reduce the price of gas for American consumers.

In: Oil Prices   ISBN: 978-1-61942-485-2
Editors: J. C. Mullen and B. M. Lynn   © 2012 Nova Science Publishers, Inc.

Chapter 1

SPECULATION, FUNDAMENTALS, AND OIL PRICES[*]

Mark Jickling, Rena S. Miller
and Neelesh Nerurkar

SUMMARY

High oil prices affect nearly every household and business in the United States. During the course of 2008, oil prices doubled to more than $145 per barrel and then fell by 80%. In early 2011, there was a run-up of about 20%, sending gasoline prices to near 2008 highs. Few would rule out the possibility of similar price swings in the months to come. What explains oil price volatility?

Some consider price movements such as those of 2008 and early 2011 to be more extreme than warranted by the fundamentals of supply and demand. Their explanation for unstable commodity prices focuses on financial markets for derivatives contracts linked to the price of oil—futures, options, and swaps. Many market participants are pure financial speculators, who never deal in physical oil, but earn large profits if they can correctly forecast price trends. Critics claim that such traders can drive oil prices above fundamental levels, resulting in a "speculative premium" that imposes unjustified costs on consumers. Although the

---

[*] This is an edited, reformatted and augmented version of a Congressional Research Service publication, CRS Report for Congress R41986, from www.crs.gov, dated September 2, 2011.

relationship between speculation and commodity prices has been studied extensively, consensus has not emerged as to whether speculative trading causes unusual oil price volatility.

An examination of Commodity Futures Trading Commission (CFTC) data reveals a strong correlation between weekly changes in positions held by "money managers" (a category of speculators that includes hedge funds) and weekly changes in the price of oil. Price falls, conversely, have tended to coincide with reductions in money managers' long positions. This statistical relationship is weaker for other classes of speculators and for commercial hedgers. However, the existence of a correlation does not imply causation—money managers could be price-followers rather than price-setters.

Another explanation for oil price volatility looks to the fundamentals of oil production and energy consumption. Rapid global economic growth led to rising demand for oil, and supply could not keep up at previous oil prices.

Because oil supply and demand do not respond much to price changes, at least in the short-term, some argue that relatively small changes in supply or demand can trigger significant price movements. An interagency task force led by the CFTC found that the 2003-2008 increase in oil prices was largely due to fundamental supply and demand factors.

The role of speculators in oil and other commodity markets has attracted congressional interest. Staff reports by the Permanent Subcommittee on Investigations of the Senate Committee on Homeland Security and Government Affairs found that excessive speculation has had "undue" influence on wheat price movements and in the natural gas market. A 2011 report by the minority staff of the House Committee on Oversight and Government Reform argues that "addressing excessive speculation offers the single most significant opportunity to reduce the price of gas for American consumers."

Legislation before the 112th Congress (S. 1200 and H.R. 2328) would authorize and direct the CFTC to take certain actions to reduce the volume of speculation in oil and related energy commodities. Another bill, H.R. 2003, would impose a tax on oil futures, swaps, and options that were not used for hedging commercial risk.

This report provides background on financial speculation in oil, the workings of oil derivatives markets, and the different types of firms that trade in those markets. It reviews the concepts of manipulation and excessive speculation, and it briefly describes the fundamental factors that affect oil prices.

INTRODUCTION

High oil prices affect nearly every household and business in the United States. *Figure 1* illustrates that during the course of 2008, oil prices doubled to more than $145 per barrel and then rapidly fell by 80%. In early 2011, there was a run-up of about 20%, sending gasoline prices near 2008 highs. Few would rule out the possibility of similar price swings in the months to come. What explains oil price volatility?

There are two possible kinds of explanations. The first looks to the fundamentals of oil production and energy consumption. Rapid global economic growth has led to rising demand for oil, and supply could not keep up at previous oil prices. But oil supply and demand are inelastic to price changes, at least in the near-term, which some would argue means that relatively small shifts in supply or demand can be expected to trigger significant price movements.

Source: Global Financial Data.

Figure 1. Crude Oil Prices: 2000-2011; (West Texas Intermediate Crude, Weekly Data).

Others consider price movements such as those of 2008 and early 2011 to be more extreme than warranted by the fundamentals of supply and demand. The second explanation for unstable commodity prices focuses on financial markets for derivatives contracts that are linked to the price of oil—futures, options, and swaps. Many market participants are pure financial speculators, who never deal in physical oil, but seek to profit from correctly forecasting price trends. Critics claim that speculators can drive oil prices above fundamental levels, resulting in a "speculative premium" that imposes unjustified costs on consumers.

Although the relationship between speculation and commodity prices has been studied extensively, there is no consensus among academics and regulators as to whether speculative trading causes episodes of unusual price volatility. This report provides background on the oil derivatives markets and the different types of firms that trade in those markets. It reviews the concepts of manipulation and excessive speculation. It includes a brief section describing the fundamental factors that affect oil prices. Although the basic question of fundamentals versus speculation remains unsettled, this report provides a context for evaluating the opposing claims.

OIL MARKETS, PRICES, AND DERIVATIVES

The United States consumes about 19 million barrels of oil each day.[1] Maintaining the supply of oil involves thousands of daily transactions, at prices varying by location, quality, quantity, and local supply and demand conditions. There is no single price at which spot (or physical) oil deals take place, but there are a number of benchmarks that buyers and sellers use as reference points. Private and government sources publish benchmark data about spot prices at various locations.

Another important oil price benchmark is the futures price. A futures contract is a form of oil "derivative"—it is a financial instrument that gains and loses value as the price of oil rises and falls. In effect, futures traders buy the price of oil (and make or lose money as the price changes) without necessarily delivering or taking possession of a single barrel of the physical commodity. Thousands of traders buy and sell oil futures contracts. Their purposes, strategies, and investment horizons vary, but the sum of their transactions determines the futures market price, which is publicly available to all market participants. Many spot trades take place at the futures price, or at the futures price adjusted by some factor.[2] Headlines reporting a dramatic jump or fall in oil prices are likely to quote the futures price.

Both physical and derivative trades (futures, options, and swap contracts linked to the price of oil) contribute to setting the price. It is thus very difficult to disentangle the price impact of trades by producers and commercial users of oil from those of financial speculators who seek to profit by forecasting price trends. Does excessive speculation drive prices away from levels justified by supply and demand fundamentals, or do speculators provide liquidity and facilitate the price-setting mechanism? These questions remain controversial. The next sections of this report describe the mechanics of oil futures and the kinds of traders who participate in the market. Although oil swaps and options use different terminology, the economic substance is the same: they are bilateral contracts under which one party gains if the price moves in one direction, and the other party gains if the price moves in the opposite direction.[3]

The Mechanics of a Futures Contract

An oil futures contract represents 1,000 barrels of oil, but neither party to the contract need ever possess the actual commodity. (Contracts may be settled by physical delivery, but in practice the vast majority are settled in cash.) When a futures contract is made today, one party (called the "long") agrees to buy oil at a future date from the other (the "short").

Contracts are available with different maturities, designated by various expiration months, but the size is always the same. (In crude oil, a contract expires every month and most trading is in the contract soonest to expire, called the "near" or "spot-month" contract.) The price at which the future purchase or sale is to take place is the current futures market price, which varies continuously during the trading day.[4] Assuming the price of oil is $100 per barrel, the long trader is committed to buy at that price, and the short is obliged to sell.

Now suppose that tomorrow the price of oil goes to $105/barrel. The long trader now has the advantage: he is entitled to buy for $100 oil that is now worth $105. His profit is $5,000 (the $5 per barrel increase times the 1,000 barrels specified in the contract). The short has lost the identical amount: she is obliged to sell oil for less than the going price.

If, on the following day, the price goes to $110, the long gains another $5,000. The short, down a total of $10,000, may reconsider her investment strategy and decide to exit the market. She can do this at any time by entering into an offsetting, or opposite transaction. That is, she purchases a long contract with the same expiration date. Her obligation (on paper) is now to sell 1,000 barrels (according to the first contract) and to buy 1,000 barrels (the second contract) when both contracts expire simultaneously. Whatever price prevails at that time, the net effect of the two transactions will be zero. The short's position is said to be "evened out"—she is out of the market.

The short's decision to exit does not affect the long, who may prefer to ride the trend. This is because all contracts are assumed by the exchange's clearing house, which becomes the opposite party on each trade, and guarantees payment. The ability to enter and exit the market by offset, without having to make or take delivery of the physical commodity, permits trading strategies based on short-term price expectations. While some traders may keep a long or short position open for weeks or months, others buy and sell in fractions of seconds.

The exchange clearing house, which guarantees all trades, also controls traders' funds. Before entering into the trade described above, both long and short would have been required to deposit an initial margin payment of $6,750. (The figure, set by the exchange, was the CME Group's margin for speculators as of August 31, 2011.) All contracts are priced, or "marked-to-market," each day. The long trader above would have had his $10,000 gain credited to his margin account, whereas the short would have had to make additional "maintenance" margin payments to cover her losses. It is worth noting that her two-day $10,000 loss represents more than 100% of her original investment, that is, her initial margin deposit of $6,750: the risks of futures speculation are high. When traders exit the market, any funds remaining in their margin accounts are returned. (Other transaction costs, such as brokerage commissions and clearing fees, are not returnable.)

Options on futures are also available for many futures contracts. The holder of an option has the right (but not the obligation) to enter into a long or short futures contract over the life of the option. The option will only be exercised if price movements are favorable to the option buyer, that is, if the underlying futures contract would be profitable. The seller of the option receives a payment (called a premium) for granting this right. The seller profits if the option is not exercised by the buyer.

Swap contracts are traded in the over-the-counter market, rather than on organized exchanges (although the Dodd-Frank Act Reform and Consumer Protection Act, P.L. 111-203, mandates that some swaps be traded on new "swap execution facilities," or SEFs). The terms of swap contracts are not uniform, as futures contracts are, but can be negotiated between the counterparties. Economically, however, swaps are equivalent to futures: one counterparty will gain if the price rises, the other if prices fall.

Who Trades in Oil Derivatives Markets

Derivatives traders can be classed as either hedgers, who use the market to avoid price risk, or speculators, who assume risk in search of profits. In futures markets the distinction is formal, and is important because hedgers pay lower margin rates and are not subject to limits on the size of their positions. Hedgers and speculators may be further broken down into subcategories, as follows.

Commercial Hedgers
Commercial hedgers are those involved in production, processing, transportation, or use of oil and petroleum products.[5] Oil and gas exploration and production companies, refiners, industrial consumers, and retailers buy and sell oil and oil products to meet the physical needs of their businesses. In their physical trading, they buy and sell oil up and down the supply chain. For example, an upstream producer sells crude oil to a refinery, which sells jet fuel to an airline or gasoline to a retail station, which then sells it to motorists. Commercial participants can sign long-term sales agreements or may buy short-term contracts for near-term physical delivery of oil.

Derivatives allow commercial participants to manage their risks related to the oil business, or hedge against oil price risk. This is a form of insurance against market fluctuations. For instance, an airline's profits may suffer when jet fuel prices increase. To address this risk, the airline can purchase long derivatives contracts whose value rises when oil or jet fuel prices increase. If prices then do increase, the cost of higher-priced fuel is offset by the money gained on derivative contracts. Alternatively, an upstream oil company can obtain a short derivative to insure against lower future oil prices.

Swap Dealers
Swap dealers are entities that deal primarily in swaps for a commodity and use the futures markets to hedge risk associated with those swap transactions. For example, a pension fund wishing to include commodities in its investment portfolio might enter into a swap linked to a published commodity price index. If the index goes up, the dealer will owe the pension fund money. To hedge

that risk, the swap dealer may take an equivalent (but opposite) position in the futures market. Then any payment due to the swap counterparty will be offset by earnings on the futures position.

A swap dealer's counterparties may be speculative traders, like pension funds or hedge funds, or producers and commercial users that are hedging risks of dealings in the physical commodity. (Many hedgers prefer swaps to futures because swaps can be customized to fit the exact quantities and time frames relevant to the hedger's business, whereas futures have uniform contract sizes and expiration dates.) Thus, swap dealer positions represent both hedging and speculation.

Money Managers

Money managers, a group of purely speculative traders, are professionally managed funds trading on behalf of clients. Money managers who invest in futures generally must register with the CFTC as commodity trading advisors (CTAs) or commodity pool operators (CPOs). The money manager class includes hedge funds, which invest on behalf of institutional investors (such as pension funds) and wealthy individuals.

Other Speculators

Other kinds of speculators include floor traders, or exchange members who trade for their own accounts, as well as a variety of firms and wealthy individuals. Small, public investors are able to trade futures,[6] but the retail presence in futures is likely much lower than in the stock market.[7]

The industry uses different terms for speculators with different time horizons. "Scalpers" take very short-term positions: minutes, seconds, or less. High-frequency trading, where the relevant time unit is the microsecond, is making inroads into derivatives trading, just as it has in the stock market. "Day traders" may hold contracts for longer intervals, but they liquidate their positions before the close of trading, to avoid exposure to overnight price risk. (As a global commodity, oil trades around the clock.) "Trend followers" may hold positions open for days, weeks, or longer, attempting to profit based on their expectations of long-term price trends.

SPECULATION AND HEDGING IN OIL FUTURES MARKETS

There are no public data on how much oil futures trading is speculative, although the assumption is that speculators account for most short-term trading, which in turn accounts for most market turnover. The Commodity Futures Trading Commission (CFTC), however, publishes weekly Commitments of Traders (COT) reports, which present data on the size of positions held by several kinds of market participants. COT data, usually published late afternoon each Friday, reflect the open interest, or the number of contracts outstanding, as of close of trading on the previous Tuesday. Thus, comparing week-to-week COT figures shows whether classes of traders have increased or decreased the size of their long, short, or spread positions.[8] The COT figures do not show how much trading has gone on during the week, or whether a position has been liquidated and then built back up, but simply offer a snapshot of positions at the market close on Tuesday.

Another significant limitation of COT data is that they do not cover swap contracts—another form of oil derivative contract not traded on exchanges. Thus, COT figures arguably cover only a subset of oil derivatives, all of which play a role in setting prices. The Dodd-Frank Act (P.L. 111- 203) gave the CFTC new regulatory authority over the swaps market. In the future, COT reports may reflect swap positions, but the data currently available cover only exchange-traded futures and options on futures.

Table 1 breaks down open interest in crude oil futures and options on futures as of July 19, 2011.[9] The figures represent the sum of identical contracts traded on the New York Mercantile Exchange (or Nymex, part of CME Group located in New York) and ICE Futures Europe (based in London). Both contracts reference West Texas Intermediate (WTI) crude, also called "light, sweet" oil, as traded in Cushing, Oklahoma (a major pipeline junction).

The data in Table 1 prompt several observations about the market:

- Speculators appear to hold most of the open interest in crude oil contracts. Producer/merchant hedgers account for only 19%; only part of swap dealers' 38% represents hedging interests;[10] and non-reportable contracts (which may be either speculative or hedges) are less than 4%. The remainder is held by speculators.
- No class of trader has a clearly dominant market share, either long or short.
- Over half of all contracts are part of spread positions, involving simultaneous purchases of long and short contracts (with different expiration months). Spread positions are less risky and offer less potential reward than outright short or long positions.[11]

- Managed money positions, which include hedge fund investments, account for a fairly small share of total open interest, but are heavily concentrated on the long side. This means that they profit when prices rise.
- Reportable positions—those with at least 350 contracts—account for more than 95% of all open interest. This suggests that small, retail investors play a minor role in oil futures markets.
- The number of reporting traders in each category is fairly small, compared (for example) with stock and bond markets, where many thousands of individuals and institutions have significant positions.[12]

Table 1. Composition of Crude Oil Open Interest: July 19, 2011 (Futures and Options Outstanding on Nymex and ICE Futures Europe)

Type of Trader	Number of Contracts	Percentage of Total	Number of Traders with Reportable Positions	
			Nymex	ICE
Producer/Merchant—Long	490,454	7.8	51	35
Producer/Merchant—Short	694,369	11.0	58	34
Swap Dealers—Long	278,753	4.4	21	13
Swap Dealers—Short	354,506	5.6	26	19
Swap Dealers—Spread	1,750,888	27.9	41	28
Managed Money—Long	278,744	4.4	90	15
Managed Money—Short	78,247	1.2	39	10
Managed Money—Spread	695,118	11.1	87	15
Other—Long	115,524	1.8	83	22
Other—Short	74,211	1.2	40	7
Other—Spread	1,236,432	19.7	107	27
Non-Reportable—Long	138,229	2.2	NA	NA
Non-Reportable—Short	100,373	1.6	NA	NA

Source: CFTC, *Commitments of Traders* report.

Notes: Figures are based on large, "reportable" positions of over 350 contracts, which must be reported daily to the CFTC. Smaller positions are combined in the "Non-Reportable" category, which includes all types of market participants.

PRICE IMPACT OF SPECULATION

In June 2011, the CFTC released a one-time report on large trader net position changes in Nymex crude oil futures, supplementing the COT reports.[13] This data set covers the period between January 2009 and May 2011, and it shows (on a weekly basis) the daily average of net position changes for both long and short positions for each of the categories of traders shown in Table 1. The figures show the amount by which long traders increased their long positions (net buys) and the amount by which short positions were increased (net sells).

Thus, for each week and for class of trader, the data show whether on average long positions (buys) exceeded short position increases (sells), or the reverse. Figure 2 presents (1) the net figure of buys and sells for managed money trading, which includes trades of hedge funds, commodity pool operators, and others; and (2) changes in the price of oil during the same period. Each point in the graph represents a single week's change in these two figures: the net average sales or purchases by money managers and the price change over the same week.

The horizontal and vertical axes divide Figure 2 into quadrants. Data points located in the upper right indicate weeks when money managers were net buyers *and* the price of oil rose. Points in the lower left indicate weeks when the price dropped and money mangers were net sellers. The other two quadrants indicate weeks when prices rose and money managers sold or when prices fell and they were net buyers: in other words, when their transactions and the price moved in opposite directions.

Figure 2 suggests that there is a correlation between money manger transactions and price movements.[14] Weeks in which the price rose sharply tended to be when they bought heavily. The more prices fell, the more they tended to sell. Very few data points fell into the upper left quadrant, that is, money managers were rarely net buyers when prices were falling.

Indeed, the results of regression analysis, given in Appendix A, show that a strong and statistically significant correlation does exist between money manger transactions and price movements. (Please see Appendix A for details of the regression.)

Figure 3 shows the same data for the trades of commercial hedgers (the group called "producer/merchants" in Table 1). Here, there appears to be no correlation, or trend-line. Neither is there any apparent correlation between the trades of (1) swap dealers or (2) other speculators and price movements, as shown in Figure 4 and Figure 5.

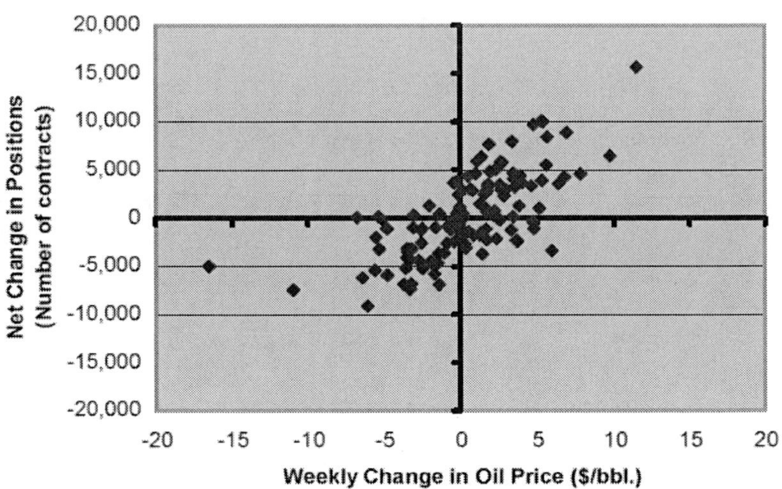

Source: Commodity Futures Trading Commission, "Large Trader Net Position Changes."

Figure 2. Net Change in Managed Money Positions and the Price of Crude Oil, (Weekly data, Jan. 2009 through May 2011).

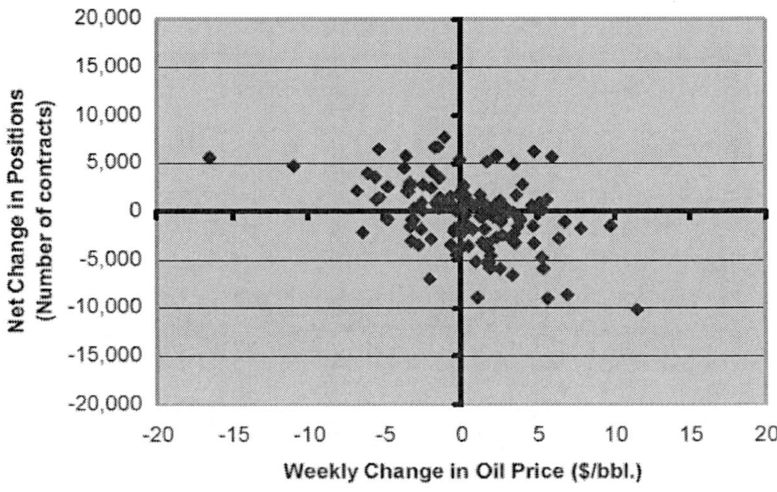

Source: Commodity Futures Trading Commission, "Large Trader Net Position Changes."

Figure 3. Net Change in Commercial Hedger Positions and the Price of Crude Oil, (Weekly data, Jan. 2009 through May 2011).

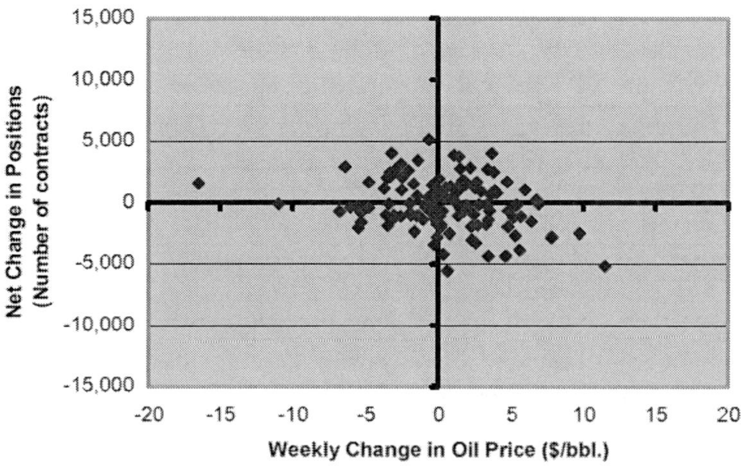

Source: Commodity Futures Trading Commission, "Large Trader Net Position Changes."

Figure 4. Net Change in Swap Dealer Positions and the Price of Crude Oil (Weekly data, Jan. 2009 through May 2011).

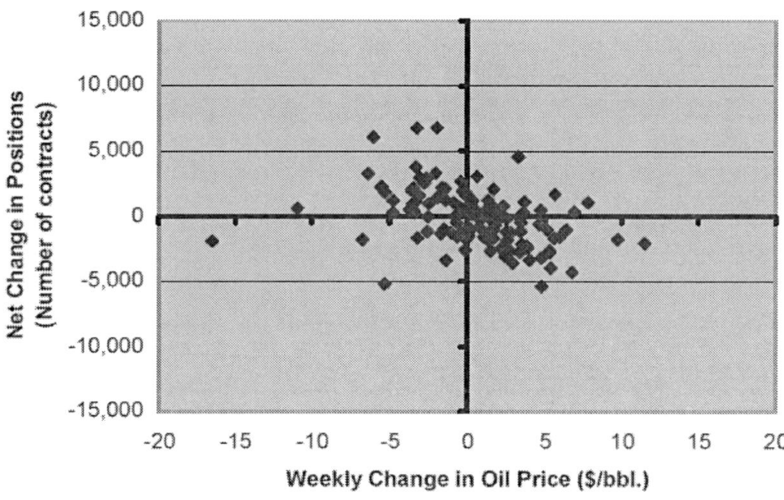

Source: Commodity Futures Trading Commission, "Large Trader Net Position Changes."

Figure 5. Net Change in "Other Reportable" Positions and the Price of Crude Oil (Weekly data, Jan. 2009 through May 2011).

Figure 2 suggests that crude oil futures are not a textbook case of an efficient market, where prices incorporating all known information about the commodity move in a random walk. The group of speculators classified as money managers appears able either to anticipate price movements or to cause those price movements through their trades. This observation raises some interesting questions. Why should money managers be better forecasters of oil price movements than other speculators or commercial hedgers? Given that their long and short positions constitute a small share of the total market, why should money manager trades have a unique price impact? Most fundamentally, are money managers' trades determining prices or are they simply more adept than others in following trends or identifying information and news that will drive prices up or down?

Assuming that money managers have a unique impact on price, what is the mechanism by which their transactions—relatively small in terms of the total market—move prices? A possibility is that they affect intraday trading, which the available open interest data fail to capture. Short-term traders might observe and seek to copy the strategies of certain money managers who are regarded as especially capable of identifying new information that might be expected to move prices, or who simply have achieved superior returns in the past.

If significant numbers of short-term speculators copy money manager trades, the impact of those trades on prices would be magnified. In effect, under this scenario, money managers may have market power beyond what the size of their positions would suggest.[15] If managed money trades trigger a significant number of similar transactions by others, they become a kind of self-fulfilling prophecy.

Such "herding behavior" among speculators, if it exists, would support arguments that the oil price at times includes a "speculative premium" above and beyond the price justified by the fundamentals.

On the other hand, it may be that money managers do trade on fundamental information and that they are especially skilled at identifying news that is going to move prices. If money managers are consistent in their ability to identify new and relevant information that will affect prices (and trade on that information before others do), one result would be the observed correlation.

A potential objection to this explanation is that it implies that some financial speculators are better analysts of the oil market than actual producers and end-users of oil, who also trade in the futures market.

Money managers might also profit by following price trends. Rather than cause price changes, they may buy when they see prices are rising and sell when prices begin to fall. But why would money managers' trading patterns, and not those of other market participants, be correlated with price changes in this way?

Other market participants may have longer investment time horizons or be less sensitive to price changes. Hedgers, for example, are generally less affected by price changes, because whatever they may lose on their futures positions, they make back in the spot market (because, for example, the physical commodity they produce will have gone up in price). Similarly, swap dealer positions may reflect long-term commodity index investments by pension funds and other institutional investors who are seeking to allocate part of their portfolio to an asset class that is not correlated to other assets they hold, such as stocks and bonds. Because the object of such investment is portfolio diversification, such investors are less likely to buy or sell in reaction to short-term price movements.

Hedge funds, by contrast, are known for taking aggressive positions in search of high yields and for seeking to extract the maximum return from any price trend. A 2008 CFTC study referred to speculators "who take positions based on price expectations over a period of days, weeks, or months" as "trend followers."[16] Trading with this time horizon would be captured by the weekly COT reports and the net position change data in Figures 2 through 5, and may be more common with money managers than other traders in oil futures and options.

If money manager trades can be said to cause price movements (that is, if we assume that such trades cause price changes, rather than follow them), are they responsible for long-term price changes such as the run-up of prices in the first half of 2008? The data released by the CFTC do not support that conclusion. When weekly position changes are plotted against changes in price in the *following* week (instead of the same week, as in Figures 2 though 5), the correlation essentially disappears. In other words, managed money trades may cause prices to rise or fall in the week they are made, but they do not appear to trigger longer price trends.

The same is true over other time horizons. For example, Figure 6 shows changes in money manager positions and price changes four weeks later. The data suggest that there is no correlation between whether hedge funds and other money managers buy or sell this week and what happens to prices over the next month.

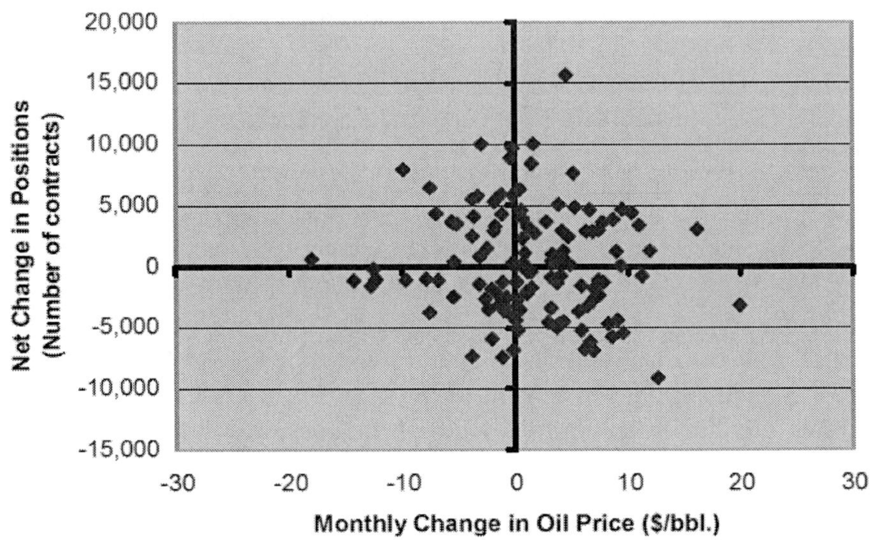

Source: Commodity Futures Trading Commission, "Large Trader Net Position Changes."

Figure 6. Net Weekly Change in Managed Money Positions and the Price of Crude Oil Four Weeks Later; (Weekly data, Jan. 2009 through May 2011).

DERIVATIVES MARKETS AND PRICE DISTORTIONS

If derivatives speculators have mispriced oil during recent years, there are two ways this could have happened. The first is through deliberate manipulation of the price by a group of market participants. Knowing action to create artificial prices is a violation of the Commodity Exchange Act, and the regulators and exchanges have market surveillance programs to detect attempted manipulation. The second possibility is much harder to evaluate: do derivatives markets *in their normal operation* have the potential to distort prices? Section 4a(a) of the Commodity Exchange Act describes "excessive speculation" as "an undue and unnecessary burden on interstate commerce," but there is no statutory or regulatory definition of the term. There is an extensive literature on the relationship between speculation and commodity prices, but the question is not settled—the data are subject to conflicting interpretations.

Manipulation

The Commodity Exchange Act (CEA), the statute which governs the regulation of commodities and futures markets, makes it a felony to manipulate or attempt to manipulate the price of a commodity, including one for future delivery.[17] Yet nowhere in the CEA does there exist a definition of the term "market manipulation." Instead, details of what it means to manipulate a futures or commodity market, in practice, have been fleshed out over the years through court decisions and regulatory actions by the CFTC.

Courts have determined that manipulation must include the following three elements:

- at the time of the alleged manipulation, there was an "artificial"[18] or "abnormal" price in the futures market;
- at the time such artificial price existed, the alleged manipulator had a dominant enough market position to permit the conclusion that he caused the artificial price; and
- the alleged manipulator intended to cause the artificial price.[19]

Section 753 of the Dodd-Frank Act (P.L. 111-203) provides the CFTC with additional anti-manipulation authority—providing false reports concerning market information that could affect prices is made a violation; anti-manipulation provisions are applied to the swap market; and criminal penalties are increased.

On April 21, 2011, President Obama announced that the Attorney General was assembling a team to root out any fraud and manipulation in the oil markets that might be contributing to higher U.S. gasoline prices. The team includes representatives from the CFTC, the Federal Trade Commission (FTC), the Federal Reserve Board, the Securities and Exchange Commission (SEC), as well as the Departments of Agriculture, Energy, Justice and Treasury.[20]

In May 2011, the CFTC filed an enforcement action against three energy trading firms and two individuals, charging them with a series of manipulations between January and April 2008 (a period when prices were rising rapidly).[21] To execute the scheme, the defendants are alleged to have created an artificial shortage of deliverable oil in Cushing, Oklahoma[22] on several occasions when Nymex futures contracts were due to expire. They did so by buying large quantities of physical oil, causing market participants to revise downwards their estimates of the amount of oil available to settle maturing futures contracts. This perception of limited available supply drove up the price, allegedly earning the defendants profits from a long position in futures. Then, having taken profits from the long position, the traders liquidated their physical oil holdings very rapidly, depressing the price and allowing them to profit from a short position in futures.

According to the CFTC, the defendants lost about $15 million on the physical side of these transactions, but earned about $50 million on the derivatives side. Their futures positions were calendar spreads—a long contract for one month and a short contract for another month. The CFTC alleges that the profitability of the scheme depended on the defendants being able to manipulate the price differentials between the two contracts in the February/March and March/April 2008 spreads. According to the complaint, the spread between the February 2008 and March 2008 Nymex futures contracts widened from $0.24 on January 3, 2008, to $0.64 on January 22, 2008.[23]

The CFTC complaint does not state how much or even whether the alleged scheme affected the price of oil itself—since the defendants were trading spreads, the absolute price level was less important to them than the difference between the various month contracts. According to the CFTC, the manipulation was put in motion on January 3, 2008, and ended on April 17, when the defendants received a request for documents from the CFTC. During that period, the price of oil rose from $99.17 to $114.80. Price increases accelerated after the manipulation ended: crude oil rose by $15 per barrel in May alone, and by another $12 in June.

Thus, although manipulations do occur in futures markets, this case (even if all allegations are proven) appears to involve short-term price dislocations that do not explain the price run-up in 2008, as it continued after the alleged manipulation ended.

Speculation

Theories of Speculation

There are two opposing theoretical views on speculation. The first is that it tends to stabilize prices and make the price-setting mechanism more efficient; the second is that at times speculation causes price trends that cannot be explained by fundamental economic factors.

The view of speculation as a stabilizing force is expressed as follows by Milton Friedman:

> People who argue that speculation is generally destabilizing seldom realize that this is largely equivalent to saying that speculators lose money, since speculation can be destabilizing in general only if speculators on the average sell when the [commodity] is low in price and buy when it is high.[24]

To make money, speculators must be able to buy low and sell high. By so doing, they smooth out the peaks and troughs of commodity prices. If they are unable to do this successfully, if their price forecasts are more often wrong than not, they will be driven out of the market by their losses.

This benevolent view of speculation is generally supported by empirical research into the effects of futures trading on cash market prices. Although the issue cannot be regarded as settled, numerous studies have found that the existence of a futures market either has no effect on or tends to reduce price volatility in the underlying commodity. For example, a recent Federal Reserve study compared price movements over the period 1991-2008 in industrial metals for which there is a futures market and metals for which no futures contract exists.[25] The study found that prices for "traded and non-traded metals are positively correlated" and that "the intensity of speculative activity in the futures markets has no explanatory power for metal price growth rates.... Instead, commodity spot price changes are driven by world economy activity and financial investors are merely responding to these price changes."[26]

In 2008, an Interagency Task Force formed by the CFTC studied price movements in crude oil, and reached a similar conclusion:

> The Task Force's preliminary assessment is that current oil prices and the increase in oil prices between January 2003 and June 2008 are largely due to fundamental supply and demand factors. During this same period, activity on the crude oil futures market—as measured by the number of contracts outstanding, trading activity, and the number of traders—has increased significantly. While these increases broadly coincided with the run-up in crude oil prices, the Task Force's preliminary analysis to date does not support the proposition that speculative activity has systematically driven changes in oil prices.[27]

More specifically, the report found that "changes in futures market participation by speculators have not systematically preceded price changes. On the contrary, most speculative traders typically alter their positions following price changes, suggesting that they are responding to new information—just as one would expect in an efficiently operating market."[28]

From an opposing theoretical perspective, speculation is seen as a potential source of price instability. Describing the behavior of investors, J.M. Keynes distinguishes between enterprise, the activity of forecasting the long-term yield of assets, and speculation, the activity of forecasting the psychology of the market. As speculators, he wrote, " ... we devote our intelligences to anticipating what average opinion believes average opinion to be."[29] In a market dominated by speculation of this type,

> A conventional valuation which is established as the outcome of the mass psychology of a large number of ignorant individuals is liable to change violently as the result of a sudden fluctuation of opinion due to factors which do not really make much difference to the prospective yield.[30]

More fundamentally, Keynes wrote that "when the capital development of a country becomes a by-product of the activities of a casino, the job is likely to be ill-done."[31] The negative view of commodity speculators is that they may trade on information unrelated to the fundamentals of supply and demand and, in the process, generate prices that harm consumers and volatility that creates uncertainty throughout the economy.

If speculators bring new fundamental information to the market, their trading should not only be profitable but should make the price discovery mechanism more efficient. When prices appear to diverge from economic reality, when a price bubble forms, many conclude that speculators are distorting the price-setting mechanism. There are several explanations for how price bubbles expand—irrational exuberance, positive feedback,[32] herding, and so on—but the process by which bubbles start and end remains little understood.[33]

How is it possible to know at any given moment whether speculation is playing a stabilizing role or whether markets are behaving irrationally? Which model of speculation best describes reality? The Commodity Exchange Act states that speculation may distort prices when it becomes excessive. The term "excessive speculation" does not provide a precise tool for distinguishing between beneficial and harmful speculation—"excess" is in the eye of the beholder—but it does provide a framework for analyzing the impact of speculation on the oil market.

Excessive Speculation

Section 4a(a) of the Commodity Exchange Act declares that "[e]xcessive speculation in any commodity ... causing sudden or unreasonable fluctuations or unwarranted changes in the price of such commodity, is an undue and unnecessary burden on interstate commerce in such commodity."[34] Unlike manipulation, however, excessive speculation is not a violation of law. The point at which speculation becomes excessive is left to the regulator to determine: there is no statutory definition or benchmark.

To many observers, phrases like "sudden or unreasonable fluctuations" and "unwarranted changes in the price" aptly describe the oil markets of 2008 and 2011. When oil prices are high and volatile, and there have been no dramatic supply shocks, many blame financial speculation.

Arguments that Oil Speculation has been Excessive

The case against oil speculators, or rather the case that oil speculation has become excessive, rests principally on two arguments. First, there is said to be too much speculative trading. While it is generally acknowledged that hedgers benefit from the presence of speculators willing to take on the risks that hedgers wish to avoid, the argument is made that financial traders have overwhelmed the market. Rather than simply provide liquidity to hedgers, speculators now account for the majority of contracts. As Table 1 shows, commercial hedging positions account for less than half of all crude oil contracts outstanding. The tail, in this view, is wagging the dog.

This view is supported by studies from the staff of the Permanent Subcommittee on Investigations (PSI) of the Senate Committee on Homeland Security and Government Affairs, which found that excessive speculation has had "undue" influence on wheat price movements[35] and in the natural gas market.[36] In the 2006 study of the impact of natural gas futures trading by the Amaranth hedge fund, the PSI found:

> Amaranth's trading demonstrates that excessive speculation can distort futures prices not only in the next month or two, but for many months into the future. Currently, the major focus of the CFTC and the exchanges is to prevent excessive speculation from disrupting orderly trading of a contract near the expiration of that contract. The CFTC and the exchanges need to be vigilant to ensure that traders' speculative positions in futures contracts several seasons, or even several years, in advance are not distorting prices.[37]

Also, a 2011 report by the minority staff of the House Committee on Oversight and Government Reform argues that "addressing excessive speculation offers the single most significant opportunity to reduce the price of gas for American consumers."[38] Others, such as CFTC Commissioner Bart Chilton, have contended that oil price gyrations are likely the result of speculative trading.[39] A frequent argument has been that a growing volume of investment flows from financial investors has affected prices.[40] One econometric analysis that incorporated oil supply and energy demand effects concluded that speculation did not explain increases in oil prices in the 2003-2008 period, although the study suggested that speculation may have played some role in previous oil price spikes.[41]

When financial institutions and investors as a group move funds into commodity markets, prices move. Even though increased financial speculation does not rise to the level of illegal manipulation, critics argue that the economic impact is the same.

In theory, higher volumes of speculative trading should not necessarily lead to more price volatility, if financial speculators base their trading decisions on the same factors as those of other market participants. But do they? The second part of the case that excessive speculation is destabilizing is that speculators do not trade on the fundamentals. The argument is that because financial speculators never produce or take delivery of physical oil, they bring to the market strategies and expectations that, in Keynes' phrase, "do not really make much difference to the prospective yield" of the asset. As a result, prices are subject to violent swings even though there has been no significant change in underlying supply and demand.

When oil prices are high, it is common to speak of a "speculative premium," meaning that the market price is higher than what the fundamentals of supply and demand justify, and that the excess is caused by uninformed speculation.[42]

The CEO of ExxonMobil Corporation addressed this issue in testimony before the Senate Finance Committee in May 2011. Asked what the price of oil would be if it were based solely on fundamentals of supply and demand, he replied that (in terms of the marginal cost of producing the next barrel of oil), "it's going to be somewhere in the $60 to $70 range."[43] But he also made more general comments on the role of speculation:

> [I]t is very difficult to precisely say what impact [speculation] has, and it's also very difficult to separate in the marketplace speculation and risk management, because the two are actually quite intertwined in terms

of how people manage the risk of the price of the fuel, whether they're a consumer or a producer.

I would give you just one benchmark. Immediately after the Libyan outbreak, the fighting that we have, within the next day the price of oil went up $12. Now, nothing had changed in the global supply the next day, so what was the market reacting to?

It was reacting to some level of insecurity about what the future supply was going to be. So that is people pricing into the global market what they believed their cost is going to be sometime in the future, building in their concerns and their worries about other possible supply disruptions and the ability of the market to respond to that.[44]

In other words, possible future supply and demand events are properly factored into today's price, even though those events may never occur. Present-day supply and demand conditions are fundamentals, but so are expectations about the future. In general, free markets are expected to distinguish between relevant fundamental information and extraneous "noise" that causes prices to drift away from fundamental values.

The argument that speculation is distorting the oil market is based on one or both of two presumed mechanisms: (1) excessive speculation by financial traders is economically (if not legally) equivalent to price manipulation, and (2) speculators introduce unwarranted volatility by trading on information unrelated to fundamentals. The next section of this report briefly analyzes the fundamentals of the oil market and suggests that sharp swings in the price of oil do not necessarily mean that prices are not based on fundamentals.

FUNDAMENTAL FACTORS

A number of factors have contributed to higher prices for oil and other energy commodities. Rapid global economic growth led to rising demand for oil, and supply could not keep up at previous prevailing oil prices. In theory, this contributes to prices rising until some consumers no longer buy oil and producers provide more supply, putting the market in balance again. But oil supply and demand is inelastic to price changes, especially in the near-term, which means it may take a larger percentage change in prices to incentivize relatively small changes in supply or demand.

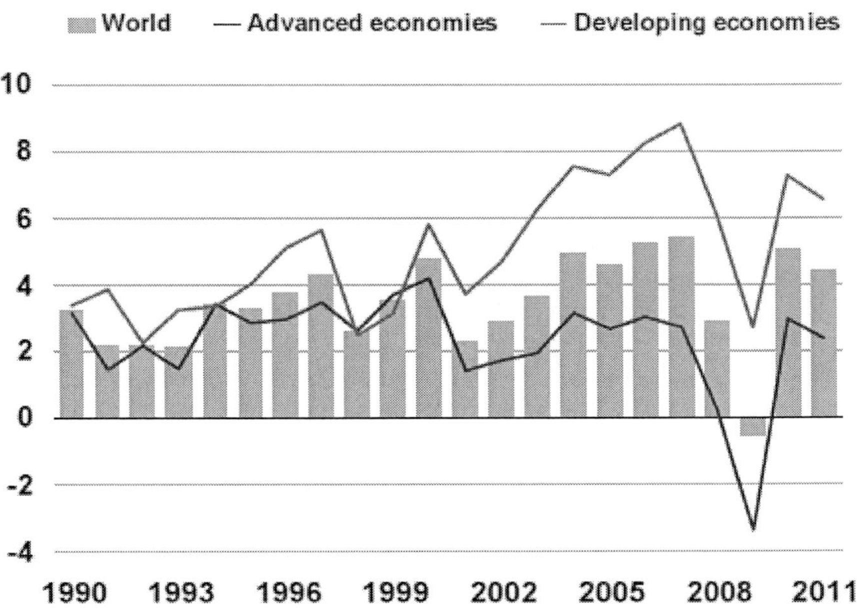

Source: International Monetary Fund, *World Economic Outlook Database,* April 2011, http://www.imf.org/ external/data.htm.

Figure 7. Global Economic Growth; (Annual growth rate of real gross domestic product (GDP), percentage).

Economic Growth Lifts Demand; Supply Cannot Keep up

Global economic growth increased demand for oil. Economic growth is the leading driver of oil demand. It accelerated to an average of nearly 5% between 2002 and 2007—the fastest five years of global economic growth on record except for the five years preceding the oil price increases of the 1970s.[45] Much of this growth took place in emerging market and developing countries, not advanced economies such as the United States or Europe. (See Figure 7.) These countries were going through the energy-intensive process of industrialization, which required greater use of energy sources such as oil and coal.

Oil supply growth, on the other hand, faced a number of hurdles. Oil resources in key oil exporting countries like Mexico, the United Kingdom, and Norway had been depleted and were in decline. Other key sources of oil supply experienced supply disruptions that reduced production. Examples included strikes in Venezuela in 2003, periodic militant attacks in Nigeria (particularly since 2003), and hurricanes in the Gulf of Mexico sometimes shutting down offshore U.S. and Mexican production.[46]

Further, some countries with abundant oil resources maintained or raised new barriers to private investment in oil exploration and production, such as increasing the national oil company's control of the energy sector, raising industry taxes, or effectively nationalizing shares in some oil producing assets.[47]

The Organization of Petroleum Exporting Countries (OPEC) played a special role among producers. This cartel of oil exporting countries varies its production level in an attempt to control oil prices. OPEC countries thus collectively maintain spare oil production capacity that can act as a cushion to oil markets in the event of disruptions or other surprising developments that require more oil. (This contrasts from production in non-OPEC countries, which is usually at maximum capacity.) OPEC did cut production at several key points when prices were falling during the 2000s. But generally OPEC members increased their output over the period when prices were rising. Perhaps more importantly, the organization ran low on spare capacity between 2004 and 2008. Low spare oil production capacity reduces the capacity to cushion against future supply disruptions or other market surprises. OPEC countries hold 77% of the world's known oil reserves, but they produce only about 42% of the world's oil supply. [48] By failing to develop more of these extensive reserves and either supply oil or establish spare production capacity, OPEC members arguably contributed to the rise in oil prices in recent years.

Price Inelasticity of Supply and Demand Enables Large Price Swings

A rising price for a good provides consumers with an incentive to consume less of that good and provides producers an incentive to supply more, which in turn can moderate the price increase. But in the oil market, supply and demand quantities can be relatively unresponsive to price changes, especially in the short run. As a result, it takes large swings in price to correspond to relatively small changes in consumption and production.

Oil is essential for economic activity and there are limited near-term substitutes. Consequently, households and industrial consumers may absorb much of the cost increases, reducing spending on other goods or reducing savings. Rapid economic growth in developing countries meant rising incomes could absorb higher energy costs. Further, some countries had subsidies that insulated consumers from the price increase during the 2000s. (Some developing countries have subsequently reformed their pricing system to reduce the fiscal burden of subsidies and reduce consumption.)

In advanced economies, incomes were already relatively high, allowing consumers to absorb higher costs for a time, albeit at the expense of other economic priorities—a painful adjustment particularly for low-income households as well as businesses and workers in industries where oil-related expenditure is a relatively significant part of the budget. Consumers could not easily reduce their consumption through efficiency improvements—the equipment that uses oil is expensive and upgrading or replacing it can require large upfront costs and take time. Examples include cars, home heating, airplanes, and industrial equipment.

These factors all contribute to global demand that is inelastic to prices: consumption did not decline in proportion to the increase in prices. In fact, except for 2008 and 2009—when the recession dragged down global demand—oil consumption has increased every year since 1993. Global oil consumption increased by roughly the same amount in the decade of the 2000s as it did during the 1990s, despite oil prices being at much higher levels. Consumption in mid-2011 has recovered from the recession, surpassing previous highs to reach record levels. Underlying these developments is a shift in consumption growth from the advanced economies to developing economies (see Figure 8).

Oil supply is also slow to respond. Oil production assets are expensive, and developing large new fields can take many years. As a result, supply can also be inelastic to prices in the short run. This tendency is exacerbated by several factors. As easy-to-develop resources have been depleted, the oil industry has moved into resources that are more complicated, expensive, and difficult to exploit, such as the oil sands in Canada or deepwater offshore developments. Where abundant easy-to-develop resources are still available, countries sometimes restrict where and how oil development and production can take place in pursuit of other national objectives, including environmental and resource management priorities. Nations may limit access to oil resources to preserve them for future generations, maintain government control of the energy industry through a national oil company, or maximize long-term revenues from energy resources.

Alternatives to oil, like biofuels, electric or gas vehicles, and coal- or gas-to-liquids technology have also faced challenges that have made them difficult to scale up. Some are expensive, require significant long term investment in infrastructure, lack sufficient technical advances, or have other potential negative impacts (consider ethanol contributing to higher food prices or coal-toliquids emitting significant greenhouse gases). Even with higher prices, many of these technologies still required public support and were thus subject to policy and political uncertainties and constraints.

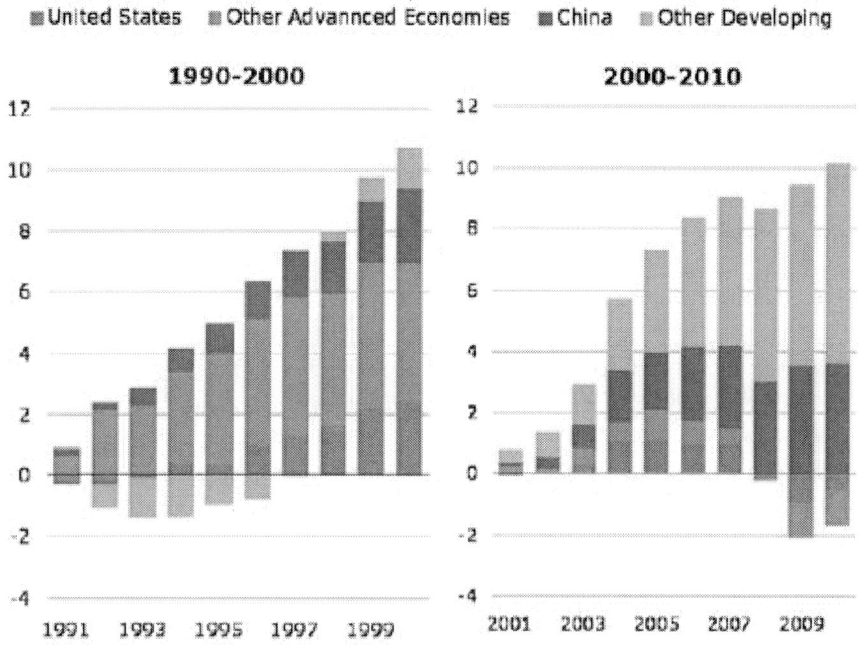

Source: EIA, *International Energy Statistics*, April 2011, http://www.eia.gov

Figure 8. Global Oil Consumption Growth: The 1990s Versus 2000s; (Cumulative demand growth over the decade, million barrels a day).

With supply and demand adjustment constrained, there was little to dampen the price increase, unlike what might take place in a more price-elastic market. However, demand is relatively elastic to income. When the U.S. economic recession spread to the world in the second half of 2008, falling incomes quickly reduced demand, and prices followed.

Again because supply is inelastic to price movements, few producers curtailed supply when prices were falling.

Even when oil prices reached near $30 per barrel, it remained above the operating cost requirements of nearly all sources of oil supply, so producers did not have to shut down for commercial reasons.

In the event, OPEC responded by making a policy decision to curtail output to support prices. These cuts reached markets in early 2009. Cuts, along with the economic recovery, contributed to subsequent price increases.

In recent months, geopolitical disruptions again came to the forefront of the market. Supply disruptions and fears of future disruptions in the Middle East and North Africa, coupled with economic recovery and the persistence of supply constraints, have contributed to oil prices reaching levels seen in the first half of 2008.

CONGRESSIONAL ACTION

Legislation before the 112th Congress (S. 1200 and H.R. 2328, both entitled End Excessive Oil Speculation Now Act of 2011) would authorize and direct the CFTC to take certain actions to reduce the volume of speculation in oil and related energy commodities.

These identical bills would impose a margin requirement of 12% for swaps and futures traded on crude oil, gasoline, diesel fuel, jet fuel, and heating oil.[49]

Such margin requirements would not apply, however, to "bona fide hedging" transactions, including those that represent a substitute for a position to be taken at a later time in a physical marketing channel, and those used to hedge a potential change in value of assets held or produced.

S. 1200 and H.R. 2328 also mandate that the CFTC impose speculative position limits on swaps and futures in crude oil, gasoline, diesel fuel, jet fuel, or heating oil that are equal to the position accountability levels or position limits established by the New York Mercantile Exchange (Nymex).

The bills include a sunset provision by which they would be terminated once the CFTC imposes position limits as required by Title VII of the Dodd-Frank Act.[50]

H.R. 2003, the Taxing Speculators Out of the Oil Market Act, would impose a tax on oil futures, swaps, and options transactions, except for those hedging commercial risk.

The tax would be levied at 0.01% of the value of a futures contract; 0.01% of the premium paid on an option; or, for a swap, 0.01% of the value of the underlying assets for each year until the swap matures.

CONCLUSION

Supply and demand issues—market fundamentals—played a central role in the increase of oil prices in recent years. Fundamental factors may have also created the conditions that enabled financial factors to have an impact on price: in a price-elastic market, purely financially driven price run-ups could quickly set off supply and demand adjustments that could again bring prices down. The absence of such adjustments may allow financial investors to drive prices for periods of time. Conditions in financial markets in energy contracts may also exacerbate volatility: relatively small changes in speculative positions appear to be associated at times with significant price swings. What remains unresolved is how much price movement in recent years is attributable to fundamental factors versus financial factors.

APPENDIX A. SPECULATIVE TRADING AND PRICE

Table A-1 shows the results of a linear regression analysis conducted using the software program STATA. The results show a statistically significant positive correlation between the net weekly change in managed money positions (i.e., the net long or short positions), and the weekly change in oil prices. The regression uses the same weekly data from the CFTC discussed in this report. The correlation between the weekly oil price change and the net positions of hedgers, swap dealers and "other" traders, however, is not statistically significant, using a 95% confidence interval.[51] In interpreting the results, the coefficient ("Coef." in column 2) shows the marginal, or incremental, effect of the independent variables on the dependent variable. Here, the independent variables are the weekly net changes in position for each of the four groups of traders; and the dependent variable is the weekly change in the price of oil. In the case of money managers, for instance, the coefficient of .7635849 implies that a net change of 1,000 positions by managed money traders would be associated with a change in the price of oil of about 76 cents per barrel. The complete dataset had 125 observations, or data points.

CRS Report R41902, *Hedge Fund Speculation and Oil Prices*, by Mark Jickling and D. Andrew Austin, finds that a similar statistical relationship holds when COT report data are analyzed, and that the correlation is robust after controlling for certain macroeconomic variables.

```
      Source |       SS           df       MS              Number of obs =     125
-------------+------------------------------              F(  5,    119) =   24.93
       Model |  946.209548        5    189.24191          Prob > F       =  0.0000
    Residual |  903.152595      119    7.5895176          R-squared      =  0.5116
-------------+------------------------------              Adj R-squared =  0.4911
       Total |  1849.36214      124    14.9142108         Root MSE       =  2.7549


   chg_price |     Coef.    Std. Err.      t      P>|t|
-------------+------------------------------------------
 swap_dealers |  .2090949   .1993362     1.05    0.296
   money_mgrs |  .7635849   .1719486     4.44    0.000
        other |  .0807596    .217542     0.37    0.711
      hedgers |  .1560825   .1664106     0.94    0.350
              |
        price |
          L1. | -.0175698   .0153826    -1.14    0.256
-------------------------------------------------------
```

Note: The coefficient on the constant term is not reported.

Figure A-1. Regression Results Using STATA Software.

APPENDIX B. OPTIONS

In the futures contracts discussed in the text, all gains by short traders create equal losses by long traders (or vice versa): futures trading is a zero-sum game. Traders who wish to limit their potential losses may choose to employ options, where gains and losses are not symmetrical. The key distinction between options and futures is that one party has the right, but not the obligation, to buy an asset in the future at a price determined when the option is purchased. There are two kinds of options: calls and puts. A call gives the holder of the options contract the right to buy an asset at a fixed price, whereas a put gives the holder the right to sell at a fixed price.

The price at which the underlying asset may be bought or sold is called the exercise price, or the strike price. An options contract confers the right to buy or sell for a specified period of time— each option has an expiration date.

On the other side of a put or call is the seller, or writer, of the option. The seller is obliged to buy or sell the asset at the strike price whenever the buyer chooses to exercise the option. In exchange for this right, the seller of the option receives a one-time payment, called the premium. The buyer's risk is limited to the amount of the premium—if prices move contrary to what the buyer expected, he simply lets the option expire unexercised, and the seller keeps the premium. On the other hand, the option buyer's potential profit is unlimited (just as a futures trader's is), because no matter how high or low the market price of the underlying asset may go, the option writer is obliged to buy or sell at the specified strike price.

The price of an option is reflected in the amount of the premium that is charged by the seller. A number of factors affect option prices: first, the relationship between the strike price and the current market price of the asset, which is called the intrinsic value of the option. If, for example, a put option on 100 shares of Company A's stock has a strike price of $14 and the current share price is $13.50, the intrinsic value of the contract to the buyer is $50 ($0.50 per share times 100 shares). An option is said to be "in the money" when the holder can exercise at a profit. If Company A's shares climbed to $15, the put option would be "out of the money," or "underwater," because the right to sell a $15 share for $14 is worthless.

In addition to intrinsic value, an option has time value. If the put on Company A in the example above is currently out of the money, there is still the chance that the share price will drop below the strike price before the option expires. Time value depends on the length of time to expiration and the price volatility of the underlying asset, which determines the probability of the option coming into the money during the life of the contract.

Options are traded both on securities and futures exchanges and over-the-counter (OTC). Underlying assets include stocks, stock indexes, futures contracts, currencies, interest rates, and physical commodities. Many OTC contracts include option-like features, including swaps, which are discussed in Appendix C.

APPENDIX C. SWAPS

Counterparties to a swap contract agree to exchange payments over a specified time period. In the simplest form, one payment is fixed, while the other fluctuates in accordance with changes in some financial variable, such as an interest rate, a stock index, or a commodity price.

In a simple oil swap, one counterparty might agree to buy 1 million barrels of crude oil from the other every three months over the next five years for $80/barrel. The other counterparty would agree to buy 1 million barrels at the Nymex spot-month price on the day the payment was due. Counterparty 1, who might be an airline wishing to hedge the risk of fuel price increases, has locked in the $80 price for five years. It will gain if the market price rises above $80.

Counterparty 2 is committed to selling oil at $80/barrel—it will benefit from the swap if current market prices drop below $80. It is possible that the second counterparty is hedging the risk of falling prices, as a producer would wish to do, but in practice most swaps involve a dealer who is willing to offer swaps on either side of the market (that is, take the floating or fixed leg of the swap). Swap dealers may manage the risk of their price exposure through another offsetting swap, or they may take an offsetting position in futures markets. (Assuming that the second counterparty in this hypothetical oil swap is a dealer, it would take a long position in futures, which would pay off if the price of oil increased. Gains in futures would offset losses on the swap; the dealer would profit through spreads and fees.)

In practice, the swapping of fixed for floating payments does not occur. The counterparties observe the change in the price of oil since the last swap payment date and calculate a single net payment, which actually changes hands.

Swaps generally do not require physical delivery of oil. Contracts in which delivery is mandatory are considered to be forward contracts and are exempt from regulation under the Commodity Exchange Act.

End Notes

[1] This figure includes crude oil and related liquid fuels such as natural gas liquids. See CRS Report R41765, *U.S. Oil Imports: Context and Considerations*, by Neelesh Nerurkar.

[2] For example, a tanker full of oil may be sold with the understanding that the buyer will pay the average futures price over the five trading days before the ship comes into port.

[3] For a description of swaps and options, see Appendices B and C.

[4] Prices are determined competitively; traders exchange bids and offers in a continuous auction process, which formerly took place on an exchange floor but now more likely involves an electronic network.

[5] The CFTC has two sets of data regarding market participants. The older set, with figures back to 1995, places banks involved in swaps dealing into the "commercial" participant category. Most swap dealers are large banks that provide over-the-counter (OTC) derivative contracts to other companies. They may use exchange-traded futures contracts to hedge the risk they take on in their OTC deals. Partly to reduce confusion about financial versus commercial participants, CFTC released a new set of "disaggregated" data with a separate category for swap dealers. The remaining commercial participants are referred to as "Producer /Merchant/Processor/Users," that is, those who are involved in the production, processing, transportation, or use of oil and petroleum products. Figures in this second data set go back to 2006. See http://www.cftc.gov/MarketReports/CommitmentsofTraders/Historical Compressed/index.htm.

[6] But not swaps, which are available only to "eligible contract participants," that is, individuals or businesses that meet asset and net worth tests.

[7] Figures on what type of trader accounts for what share of transactions are not available.

[8] In a spread position a trader has a long contract for a given month and a short contract for a different month. In effect, a spread position is a bet that the difference in prices between the futures contracts for the two months will widen or narrow. A spread position, other things equal, carries less risk than an outright long or short position, because whatever happens to the general price level, both legs of the spread will move in tandem to some degree.

[9] These percentages are relatively constant over time. See CRS Report R41902, *Hedge Fund Speculation and Oil Prices*, by Mark Jickling and D. Andrew Austin.

[10] The CFTC estimates that in May 2011 index traders, who are generally speculators, accounted for the equivalent of 669,000 contracts in crude oil, which is more than 40% of total swap dealer positions. See "Index Investment Data," available at http://www.cftc.gov/ucm/ groups/public/@marketreports/documents/file/indexinvestment0511.pdf.

[11] These figures may understate the extent of spread trading, because they do not include intermarket spreads (for example, a long position in crude oil against a short position in jet fuel, which would be a bet that the price differential between the two commodities will change, irrespective of the overall price level). Some portion of the COT long and short interest represents intermarket spreading, but the exact amount is not known.

[12] The cash margin required of a speculator holding 350 Nymex contracts is less than $2.5 million dollars, which would not be considered a particularly large position in the stock market, where daily turnover in the United States exceeds $100 billion. Also, note that the number of reporting traders on Nymex and ICE should not be summed, because the same entities may trade on both exchanges.

[13] CFTC, "CFTC Publishes Two New Data Sets on Daily Net Position Changes," Release PR6066-11, June 5, 2011, available at http://www.cftc.gov/PressRoom/PressReleases /pr6066-11.html.

[14] This correlation also appears in Commitments of Traders data and appears to hold under more advanced modeling. See CRS Report R41902, *Hedge Fund Speculation and Oil Prices*, by Mark Jickling and D. Andrew Austin.

[15] In economics, market power means the ability to alter prices. Under perfect competition, all firms are price takers; those with market power are price makers.

[16] CFTC, *Staff Report on Swap Dealers and Index Traders*, September 11, 2008, p. 39, available at http://www.cftc.gov/PressRoom/PressReleases/pr5542-08.html.

[17] Section 9(a) of the Commodity Exchange Act, 7 U.S.C. §13(a).

[18] The definition becomes somewhat circular: manipulation is that which causes artificial prices, while artificial prices are those caused by manipulation. See Craig Pirrong, "Squeezes, Corpses, and the Anti-Manipulation Provisions of the Commodity Exchange Act," *Regulation*, vol. 17, no. 4, 1994.

[19] See *Frey v. Commodity Futures Trading Commission*, 931 F. 2d 1171 (7th Cir. 1991).

[20] "Obama Team To Probe Oil Market Manipulation," Reuters, April 21, 2011, at http://www.reuters.com/article/2011/04/21/us-obama-oil-market-view-idUSTRE73 K7B420110421

[21] CFTC, *Complaint for Injunctive and Other Equitable Relief and Civil Monetary Penalties Under the Commodity Exchange Act: U.S. Commodity Futures Trading Commission v. Parnon Energy Inc., Arcadia Petroleum LTD, Arcadia Energy (Suisse) SA, Nicholas J. Wildgoose and James T. Dyer*, May 24, 2011, available at http://www.cftc.gov/ucm/ groups/public/@lrenforcementactions/documents/legalpleading/enfparnoncomplaint052411. pdf.

[22] The delivery point for Nymex WTI oil futures.

[23] CFTC, *Complaint*, p. 13.

[24] Milton Friedman, *Essays in Positive Economics* (Chicago: University of Chicago Press, 1953), p. 175.

[25] The metals studied were copper, aluminum, lead, nickel, tin, and zinc (all of which have actively traded futures contracts in New York or London) and steel, manganese, cadmium, cobalt, tungsten, rhodium, ruthenium, and molybdenum (for which there are no exchange-traded futures contracts).

[26] George M. Korniotis, "Does Speculation Affect Spot Price Levels? The Case of Metals with and without Futures Markets," Board of Governors of the Federal Reserve, Divisions of Research & Statistics and Monetary Affairs, Finance and Economics Discussion Series, Staff Working Paper 2009-29, May 2009, pp. 27-28.

[27] Interagency Task Force on Commodity Markets, *Interim Report on Crude Oil* , July 2008, p. 3. (The task force included staff from the CFTC, the Departments of Agriculture, Energy, and the Treasury, the Federal Reserve, the Federal Trade Commission, and the Securities and Exchange Commission.)

[28] Ibid.

[29] John Maynard Keynes, *The General Theory of Employment Interest and Money* (London: Macmillan and Co., 1936), chap. 12, sec. V. Keynes also describes speculative markets as beauty contests in which judges select not the contestant that they personally find most attractive, but rather the contestant that they believe a majority of spectators would select.

[30] Ibid.

[31] Ibid, sec. VI.

[32] Positive feedback is a way to describe trend-following. Speculators buy, the price goes up, more speculators buy, and the price continues to rise, regardless of fundamentals. At some point, speculators realize that prices are too high, but buy anyway, betting that they will be able to sell to a "greater fool" before the bubble bursts.

[33] Federal Reserve Chairman Ben S. Bernanke, "Implications of the Financial Crisis for Economics," speech at Princeton University, September 24, 2010, http://www.federalreserve.gov/newsevents/speech/bernanke20100924a.htm.

[34] 7 USC § 6a(a)(1). To prevent excessive speculation, Section 4a(a) directs the CFTC to impose limits on the size of positions that speculators can hold in swaps and futures markets.

[35] U.S. Senate, Committee on Homeland Security and Governmental Affairs, Permanent Subcommittee on Investigations, *Excessive Speculation in the Wheat Market*, Majority and Minority Staff Report, June 24, 2009, available at http://hsgac.senate.gov/public/_files /REPORTExcessiveSpecullationintheWheatMarketwoexhibits chartsJune2409.pdf.

[36] U.S. Senate, Committee on Homeland Security and Governmental Affairs, Permanent Subcommittee on Investigations, *Excessive Speculation in the Natural Gas Market*, Staff Report with Additional Minority Staff Views, June 25, 2007, available at http://hsgac.senate.gov/public/index.cfm?FuseAction=Subcommittees.Investigations.

[37] Ibid., p. 120.

[38] *Real Help for American Consumers: Who's Profiting at the Pump?* May 23, 2011, p. 13, available at http://democrats.oversight.house.gov/images/stories/ FULLCOM/524%20oil% 20products/ COOGR%20Democratic%20Oil%20Report%2005-23-11.pdf.

[39] CFTC Commissioner Bart Chilton, "Speculators and Commodity Prices," Futures Industry Association's Panel Discussion: Financial Investors' Impact on Commodity Prices, Boca Raton, FL, March 16, 2011, available at http://www.cftc.gov/PressRoom/Speeches Testimony/opachilton-41.html.

[40] Kenneth J. Singleton, "Investor Flows and the 2008 Boom/Bust in Oil Prices," March 23, 2011, available at http://www.stanford.edu/~kenneths/OilPub.pdf.

[41] Lutz Kilian and Dan Murphy, "The Role of Inventories and Speculative Trading in the Global Market for Crude Oil," Center for Economic and Policy Research, Discussion Papers 7753, May 2010.

[42] For example, the Saudi oil minister recently stated that "surging oil prices were primarily owing to speculations [and] baseless information and concerns over supply and demand." He argued that since supply, demand, and oil reserves were in balance, there was no reason for higher prices. "Saudi Oil Minister Blames High Oil Prices on Speculations," Kuwait News Agency, April 9, 2011.

[43] Testimony of CEO Rex Tillerson, Exxon Mobil Corp., in U.S. Congress, Senate Finance Committee hearing on "Oil and Gas Tax Incentives and Rising Energy Prices," May 12, 2011, in reply to a question from Senator Cantwell. (From Congressional Quarterly transcript.) The price of crude oil was then near $100/barrel.

[44] Ibid.

[45] Christof Rühl, "2008 BP Statistical Review of World Energy (speech)," London, June 2008, p. 2, http://www.bp.com/statisticalreview.

[46] CRS Report R41765, *U.S. Oil Imports: Context and Considerations*, by Neelesh Nerurkar.

[47] For more information, see CRS Report RL34137, *The Role of National Oil Companies in the International Oil Market*, by Robert Pirog.

[48] BP, *2011 Statistical Review of World Energy*, June, 2011, http://www.bp.com/statisticalreview.

[49] Sec. 2(b)II(C), S. 1200 and H.R. 2328, 112th Cong., 1st sess. (2011). Margin levels for commodity futures generally fluctuate and are set by exchanges on which futures are traded. Although margin levels vary, they typically range from about 2% to 10% of the full value of the futures contract—as of August 30, 2011, the margin for Nymex oil futures was 7.7%.

[50] Title VII of the Dodd-Frank Act (P.L. 111-203) authorizes the CFTC to increase margin requirements but mandates that the CFTC impose position limits on commodity derivatives such as oil swaps.

[51] To determine statistical significance, examine the "P-value" in the column headed P>|t|. When the P-value is less than 0.05—as is the case for the value 0.000 for money managers, but not for the other three categories of hedgers, swap dealers and "other" traders—then the correlation is statistically significant within a 95% confidence interval.

In: Oil Prices   ISBN: 978-1-61942-485-2
Editors: J. C. Mullen and B. M. Lynn   © 2012 Nova Science Publishers, Inc.

Chapter 2

OIL PRICE FLUCTUATIONS[*]

Neelesh Nerurkar and Mark Jickling

SUMMARY

Oil prices generally increased from 2002 until mid-2008, collapsing with the economic downturn and then rebounding with global economic recovery in 2009 and 2010. In 2011 to date, the price of West Texas Intermediate crude oil has ranged from less than $80 to more than $110. These price levels and fluctuations stand in contrast to the relatively low and stable oil prices of the 1990s, when the level of WTI averaged about $20 per barrel. Periods of price increases raise economic and energy security concerns. This report provides an overview of factors that contributed to these oil price movements and may continue to drive prices in the future.

The United States imports about half of the oil and related liquid fuels needed to meet domestic demand. The volume of imports is down in recent years due to higher domestic production and lower consumption. Nonetheless, oil import costs have climbed due to higher oil prices. Rising oil prices can strain household budgets, widen the trade deficit, and create economic dislocations that reduce economic growth and lower employment.

The oil market is globally integrated and affected by a range of international supply and demand developments (market fundamental factors). Rapid, energy-intensive economic growth in developing

---

[*] This is an edited, reformatted and augmented version of a Congressional Research Service publication, CRS Report for Congress R42024, from www.crs.gov, dated August 26, 2011.

countries raised global oil demand since 2002. Global oil supply was unable to keep up at previously prevailing prices. Depletion of some easy-to-produce resources, the oil industry's difficult shift to more complex resources, geopolitical and weather-related supply disruptions, resource nationalism in oil rich countries, and the actions of the Organization of the Petroleum Exporting Countries all hampered supply growth.

Oil prices rose to balance the market, pricing out some consumers and incentivizing additional supply. Supply and demand are inelastic to price changes, so it took large changes in prices to affect consumers and producers. However, demand is responsive to economic conditions. The recession, which caused a sharp decrease in global oil demand, proved to be a brief respite from high prices. Prices recovered as economic growth returned, particularly in developing countries, and new geopolitical supply disruptions again raised concerns about global supply. How these long- and short- term supply and demand trends develop will shape future price movements.

The rising price of oil has also raised concerns that non-fundamental factors increased prices even higher than justified by supply and demand. The inelasticity of supply and demand may have enabled other factors to further raise prices.

Of particular concern has been the impact of rising financial investment in energy derivatives, as well as concerns about impacts of exchange rate fluctuations and price manipulation. These factors may impact the price of oil and oil products to some degree for limited periods of time, but it is unclear how large or enduring their impacts are.

A number of policy measures have been introduced by Congress and the administration to address both fundamental factors and non-fundamental factors.

Fundamentals policy is complicated by the high upfront costs and long lead times that energy sector investments need to bear fruit. Much of the fundamentals focused legislation advanced in the 112th Congress focuses on the supply side of the equation. Non-fundamental factors have also been addressed, most notably as part of the Dodd-Frank Wall Street Reform Act of 2010.

However, delays in implementation mean it is still unclear what impact these measures will have. Given the scale and complexity of the challenge posed by various drivers of the oil price, it may require a suite of options to best address oil price risks. Given the long lead times for investment and technology development and deployment to provide benefits, commitments to such measures may need to persist through continued upward and downward oil price fluctuations.

INTRODUCTION

This report provides an overview of factors that affect the price of oil. The price of West Texas Intermediate crude oil (WTI), the main U.S. benchmark oil price, generally increased from 2002 until mid-2008.[1] It reached a one-day high closing price of $145 per barrel (/bbl) on July 3, 2008, then collapsed to $30/bbl by December 23, 2008.[2] Prices rebounded in 2009 and 2010 and have fluctuated from more than $110/bbl to less than $80/bbl in 2011 to date. Oil price fluctuations are of great interest to policy makers, particularly when they result in higher prices. Oil is an essential input to the economy, and oil prices are the main driver of prices for gasoline and other petroleum products (see Figure 1). A wide range of factors can contribute to oil price fluctuations, including economic, commercial, geopolitical, and financial developments.

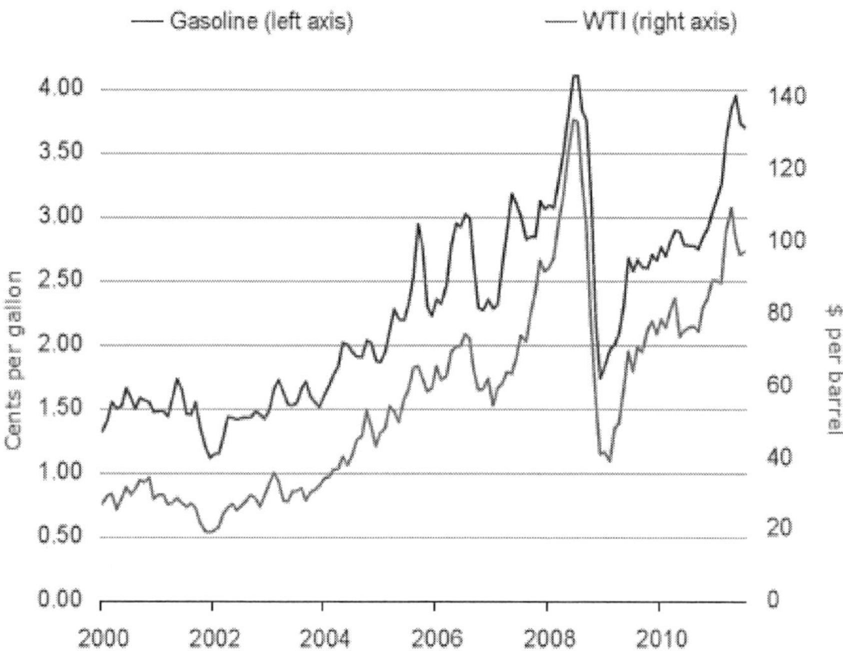

Source: EIA, http://www.eia.doe.gov/petroleum.

Figure 1. Monthly Oil and Gasoline Prices.

The U.S. Situation

The Unietd States consumes roughly 19 million barrels of oil and related liquid fuels per day (Mb/d). About 72% of oil consumption is used for transportation, 22% is used by the industrial sector, and 5% is used directly by households and commercial buildings.[3] Almost no oil is used to generate electricity in the United States. The United States produced nearly 10 Mb/d of oil and related liquid fuels in 2010. Domestic supplies included 5.5 Mb/d of crude oil production, of which 1.8 Mb/d came from offshore sources. Natural gas liquids, such as propane and butane, accounted for 2.1 Mb/d. Fuel ethanol, mostly from corn, constituted nearly another 1 Mb/d, with the rest of domestic supply from other sources.[4]

The United States imported about 9 Mb/d of oil and other liquid fuels in 2010. U.S. net imports declined 3.1 Mb/d between 2005 and 2010. The recession and high oil prices reduced oil demand during this period, and consumption fell by 1.6 Mb/d. At the same time, domestic production increased about 1.5 Mb/d, led by higher domestic output of ethanol, offshore oil from the Gulf of Mexico, and natural gas liquids. Higher prices and technology improvement also enabled supply growth from onshore sources through enhanced oil recovery and hydraulic fracturing. Of remaining demand met through net U.S. imports in 2010, roughly 20% came from members of the Organization of the Petroleum Exporting Countries (OPEC) in the Persian Gulf region, and another 30% came from OPEC members elsewhere. But Canada was the largest single foreign supplier, accounting for 24%.[5]

Despite rising production, falling consumption, and fewer barrels of oil being imported, the cost of oil imports has increased in recent years due to higher oil prices. The cost of net petroleum imports averaged $726 million per day in 2010, which is up from $628 million per day in 2005,[6] though this is lower than the $1.1 billion per day spent in 2008 when WTI averaged about $100/bbl.[7] Net petroleum import costs rose to $982 per day on average in June of 2011, when WTI averaged $96/bbl.

The United States in a Global Oil Market

The oil market is globally integrated, and developments around the world affect prices everywhere. For instance, European refiners were the primary recipients of Libyan crude prior to recent unrest which shut down Libyan oil exports. European refiners now may bid for oil from elsewhere, in the same global market place where U.S. and Asian refiners buy crude, transferring the price effect of scarcity around the globe. And, while U.S. oil consumption may be lower than it was a few years ago, global consumption is at an all time high. Oil demand abroad surged in 2010 as the global economy rebounded. Global oil consumption in the second quarter of 2011 was estimated to be 87.4 Mb/d, 1.1 Mb/d higher than in the second quarter of 2008,[8] when crude prices were climbing to their all time peak (reached on July 3, 2008, for WTI).

However, recent oil price fluctuations are not divorced from U.S. developments. For instance, concerns about U.S. economic growth contributed to falling oil prices in early August 2011.[9] As the world's largest oil consumer, developments that could impact U.S. consumption can have significant impacts on global oil prices.

Three important aspects of the oil market help explain these prices fluctuations. First, oil supply and demand are relatively unresponsive to price changes, especially in the short run (see "Elasticity and the Oil Market"). It takes relatively large changes in price to prompt relatively small changes in quantities consumed or produced. Correspondingly, small changes in consumption or production can prompt large price swings.[10] Second, oil consumption is highly responsive to economic growth, the most important factor in determining oil demand. Finally, markets are forward looking, which means concerns about future supply (e.g., if more unrest in the Middle East and North Africa disrupts oil production) and demand (e.g., if Chinese oil consumption grows quickly) affect the price at which commercial and financial market participants are willing to buy and sell oil today.

To explain the drivers behind and impacts of oil price fluctuations, this report first turns to how oil prices affect the economy. It then looks at global oil market fundamentals, supply and demand issues that contributed to oil price fluctuations and generally higher oil prices. The final section turns to the role of factors apart from supply and demand, non-fundamental issues such as how financial speculation affects oil prices.

OIL PRICES AND THE U.S. ECONOMY

Oil is critical to the U.S. economy. It is the United States' largest source of energy, providing 37% of the total energy the nation consumes and 94% of the energy used for transportation. Every U.S. recession in the last 40 years has been preceded by an increase in oil prices.[11] Some economic analysts estimate that, as a rule of thumb, the impact of a sustained $10/bbl increase in the price of oil could result in about 0.2% lower economic growth and 120,000 fewer jobs in the first year after the price increase, with impacts worsening in the second year.[12] That said, the mechanisms by which higher oil price levels and volatility work through the economy to create these negative impacts is complex.

Rising oil prices drive up the cost of petroleum products which then take up a larger part of U.S. consumers' budgets. Figure 2 illustrates how direct personal expenditures on oil have changed. Increased cost of oil products leaves less to spend on other goods and services, or to put into savings. If oil-consuming businesses can pass on their higher fuel costs to consumers, oil can indirectly raise costs on other goods and services, and potentially contribute to higher interest rates.[13] If oil-consuming businesses cannot pass on fuel costs, then higher oil costs can burden company budgets. This compounds stress from weaker sales to customers with less money to spend on their products. As a result, oil-consuming businesses may reduce investment and employment.

Rising oil prices can shift wealth from oil consumers to oil producers in the United States and abroad.[14] The portion received by domestic producers remains in the economy, but is redistributed away from consumers and non-oil businesses and towards oil companies, their shareholders, and to those that provide goods and services to the oil industry. The economy can adjust to redistribution, but the adjustment can take time and be painful. Workers in non-oil businesses may lose their jobs and have to be retrained to enter different industries; factories may be idled— causing the underutilization of previous investments and productive capacity.[15]

Because the United States imports half its oil supply, an oil price increase would also result in some additional wealth being sent abroad to foreign oil producers. The size of the U.S. oil import bill is discussed above. How these higher import costs impact the economy is in part based on how foreign oil exporters spend their rising wealth. Some of this wealth is returned to oil importing countries if wealthier oil exporters purchase more of the oil importer's goods and services. This has been much more the case for Europe and China than for the United States.[16] Many oil exporters have not tended to spend as much of their rising oil wealth on importing more U.S. goods and services.

Oil exporting countries may also purchase assets from, or lend money to, oil importing countries through financial markets. This financial inflow could help sustain a trade deficit.[17] This availability of capital from oil exporters to oil importers may also help blunt the economic dislocations that oil importing countries experience when oil prices increase.[18] However, this financial cushion comes at the expenses of asset sales or obligations to repay debts in the future.

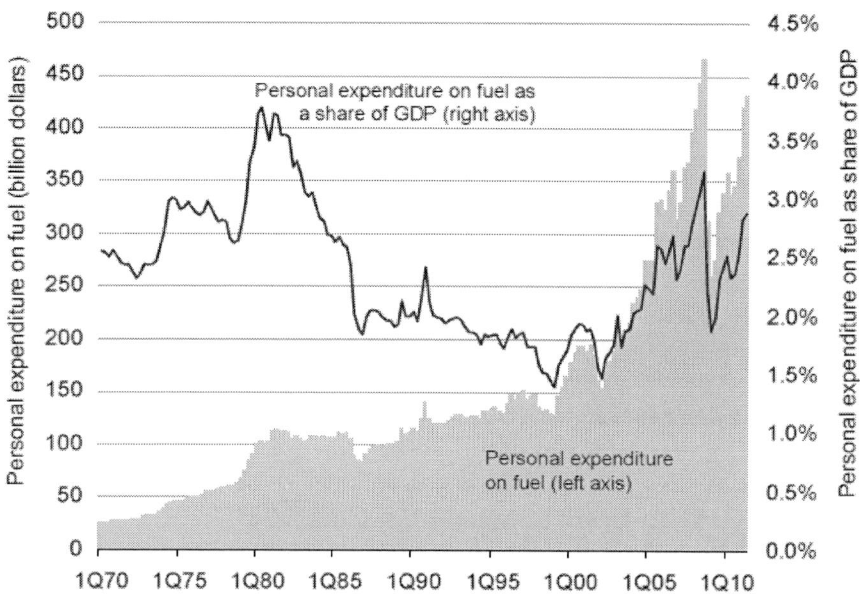

Source: U.S. Bureau of Economic Analysis, Table 2.4.5. Personal Consumption Expenditures by Type of Product, National Income and Product Accounts Tables, July 29, 2011, http://www.bea.gov/iTable/iTable.cfm? ReqID=9&step=1.

Notes: Includes direct personal consumption expenditures for motor vehicle fuels such as gasoline, heating oil, and other petroleum based fluids. Does not include higher expenditures if higher fuel costs indirectly increased the costs of other goods that households purchase (i.e., inflation caused by oil).

Figure 2. Quarterly Personal Spending on Oil, 1Q1970-2Q2011.

How and when oil price increases occur determine their economic impact. A sharp, quick increase may have larger negative impacts than gradual increases because they leave less time to adjust.[19] Price volatility may provide confusing market signals, leaving companies uncertain as to what steps to take to adjust. Also, broader economic conditions at the time of the oil price increase matter—a strong economy may see growth dampened only slightly, while an oil price increase in a weak economy may be enough to push it into recession. This is likely what occurred in 2007- 2008: rising oil prices helped turn the U.S. economic slowdown into a recession.[20]

MARKET FUNDAMENTALS: GLOBAL SUPPLY AND DEMAND

Rapid, energy-intensive economic growth in developing countries has contributed to rising global oil demand. A range of barriers to production growth kept oil supplies from being able to keep up with demand at prevailing prices. Oil prices rose to balance the market—which forced out some consumers and incentivized additional production. But oil supply and demand are relatively inelastic to price changes: Large changes in price correspond to small changes in quantities produced or consumed. Consequently, substantial price increases were necessary to slow demand growth. Oil demand is sensitive to economic growth; demand and prices fell with the global economic downturn, but rebounded with economic recovery. These fundamentals factors are likely to lead to future price fluctuations and, some believe, may contribute to long term price increases.

Demand Growth Led by Developing Countries

Global economic growth between 2002 and 2007 was rapid and energy intensive. This was the fastest five-year period of economic growth since the early 1970s—when economic growth also contributed to an increase in oil and other commodity demand and prices.[21] Economic growth in the 2000s was led by developing economies, such as China and India (see Figure 3). While economic growth in these countries slowed in 2009, it remained positive and rebounded quickly in 2010.

Economic growth translates to demand for oil and other sources of energy. Many developing nations are industrializing, a particularly energy intensive stage of economic development. Industries commonly use oil to fuel industrial processes and transport goods or as a feedstock in manufacturing. Households in these countries are also purchasing energy-intensive goods, such as heating and personal transportation. Rapid economic growth means rising incomes, which helps offset the higher cost of energy for businesses and households in emerging economies. And some of these consumers never experienced rising energy costs the way U.S. consumers did because their governments subsidized energy consumption, controlling domestic prices. The difference between domestic and international prices was paid by the government. Some major subsidizers are now removing these policies, moving toward market based pricing.[24]

Elasticity and the Oil Market

Economists use the concept of elasticity to measure the responsiveness of supply and demand to changes in prices and income. The price elasticity of demand or supply is calculated as the percentage change in quantity consumed or produced of a product that results from a specified percentage change in the product's price.

Price elasticity varies depending on the time frame chosen. In the short term, both the oil demand and supply are particularly inelastic to price. When the price of oil rises, consumers are not likely to reduce consumption by a similar amount. The reason for this behavior is that there are few available substitutes for fuels such as gasoline or home heating oil in the short term. Many consumers will continue to buy the product even if the price goes up and it can take large changes in price to get consumers to stop buying oil. Some of the easiest efficiency improvements were already carried out after the oil shocks of the 1970s. Some analysts estimate global short term price elasticity of oil demand is around -0.1, meaning that a 10% increase in oil prices would result in only a 1% decline in consumption, holding all else constant.[22] Similarly, most oil producers have little capability to increase the supply in the short term, even with the incentive of higher prices. Significant time lags are required before the discovery and development of new oil fields yield increased supply on the market.

Long term oil demand and supply elasticity to price is likely to be greater than in the short term. But price elasticity is still relatively low as it requires fuel switching, new technologies, and access to resources. Further, the large price fluctuations in oil prices create uncertainty that may undermine willingness to make these investments, reducing the long term supply and demand responsiveness to prices.

In contrast, oil demand is relatively much more elastic (responsive) to income, frequently measured as GDP. Oil is critical to the functioning of modern economies. For instance, it is necessary for fueling industrial processes, as feedstock for chemicals and other materials, transporting goods, and transporting people to their jobs. As a result, oil demand grows in lock step with many economies. Oil demand increasingly comes from uses where oil is difficult to substitute, such as transportation, and from places with high income elasticities, like in the developing world.[23]

The experience of developing countries stands in contrast to that of advanced economies in recent years. Advanced economies represented most of the oil consumption growth in the 1990s, but rising oil prices and slower economic growth contributed to falling oil consumption in these economies after 2005 (see Figure 4). However, the advanced economies are the wealthiest consumers in the world. Further, oil consumption in these countries is concentrated in transportation, where oil is particularly difficult to substitute. This makes demand inelastic to price, and some analysts have suggested it has become even more so in the United States as suburban development has led to greater driving distances between home and work.[25] It was only after the recession started in 2007, further weighing down demand, that the largest decreases in U.S. and then other advanced economies' oil consumption occurred. While demand can be unresponsive to price changes, it is responsive to changes in incomes, which fell with the recession.

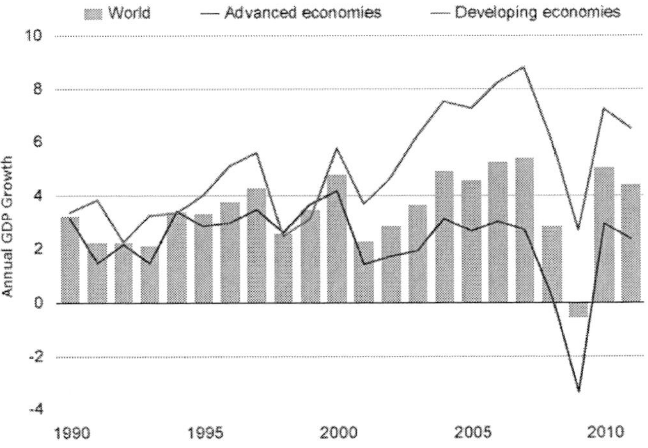

Source: International Monetary Fund, *World Economic Outlook Database,* April 2011,
        http://www.imf.org/ external/data.htm

Notes: Advanced economies include Australia, Austria, Belgium, Canada, Cyprus, Czech Republic, Denmark, Estonia, Finland, France, Germany, Greece, Hong Kong, Iceland, Ireland, Israel, Italy, Japan, Korea, Luxembourg, Malta, Netherlands, New Zealand, Norway, Portugal, Singapore, Slovak Republic, Slovenia, Spain, Sweden, Switzerland, Taiwan, United Kingdom, and United States. Developing economies includes the rest of the world.

Figure 3. Global Economic Growth; Percent growth of gross domestic product, constant prices.

Therefore, rapidly rising economic activity in developing countries boosted their oil consumption despite price increases. But, as illustrated by the dotted 'World' line in Figure 4, cumulative oil consumption growth in the 2000s was actually less than in the 1990s, when prices were relatively stable. This slow growth points to additional factors that contributed to recent oil market developments, particularly difficulties in growing supply, which is discussed next.

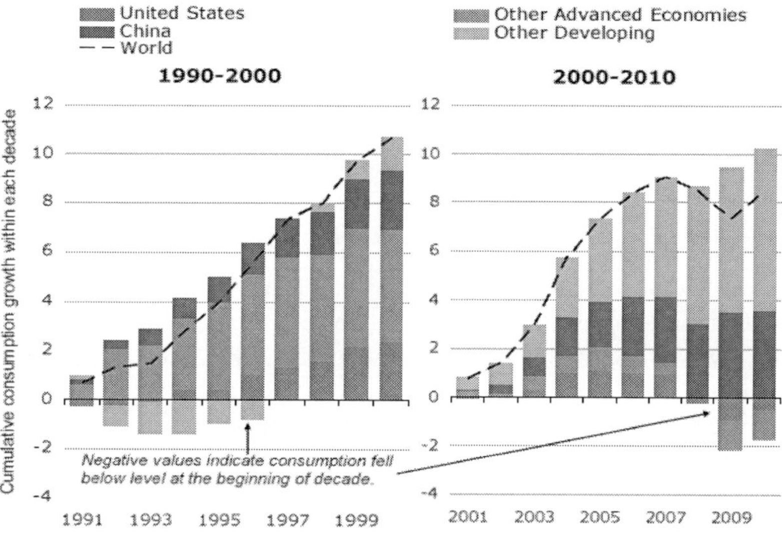

Source: EIA. See Figure 2 notes for explanation of Advanced Economies.
Notes: This chart illustrates consumption growth within each decade, i.e., the growth from 1990 to 2000 versus the growth from 2000 to 2010. Cumulative growth over the course of the 1990s was 10.7 Mb/d and cumulative growth over the 2000s was 8.5 Mb/d (net of declines in advanced economy consumption).

Figure 4. Cumulative Oil Consumption Growth: The 1990s versus the 2000s; Consumption growth from 1990 to 2000 and from 2000 to 2010 in Mb/d.

Challenges to Oil Supply Growth

Several barriers kept oil supply from increasing fast enough to stem recent price increases. Market analysts in the early 2000s had expected that oil prices much above historic levels would give rise to a robust supply response that could limit subsequent price increases.[26] Oil price increases in the 1970s and in the early 1990s had been followed by significant supply growth outside of OPEC.[27] But several factors inhibited a supply response that could have prevented subsequent price increases.

First, oil supply is coming from progressively more remote and less conventional deposits. Where accessible, the easiest oil to produce is exploited first. Historically, these tend to have been the large, conventional deposits located onshore or at shallow water depths. Oil production from several key conventional areas fell during the 2000s, including Alaska, Australia, Indonesia, Mexico, Norway, and the United Kingdom. Second, as those cheaper and easier resources become depleted, the oil industry has moved progressively toward resources that are more difficult and more expensive to recover. Production is increasing from more complex deposits of oil, such as deepwater offshore resources or oil sands, which are more expensive and can require longer lead times to develop. Third, supplies of competing alternative transportation fuels such as coal- and gas-based transport fuels, biofuels, or electric transportation, are proving more expensive and difficult to scale up than may have been previously anticipated. Thus the global transportation system continues to depend almost entirely on oil.

Supply challenges were exacerbated by shortages and cost inflation in required exploration and production equipment. Rapid economic growth was widespread in the 2000s, with many industries demanding basic materials like steel and cement. These materials are necessary for oil projects and contributed to rising project costs.[28] Also important was a shortage of specialized personnel and equipment needed for oil project development, such as experienced engineers and geologists or drilling rigs.[29] Oil field development has always been capital intensive and usually has required long lead times, but the new, larger and more complex projects can have even longer lead times and present greater logistical challenges. Industry investment has increased with the increase in prices, but it can take many years for that investment to result in new supplies and at least some of that increase in investment has been eaten up by cost inflation. Consequently, the cost of oil production increased rapidly for many producers during the 2000s[30] and projects were delayed, widening already long lead times and slowing supply growth.[31] Furthermore, those challenges were not limited to conventional and unconventional oil projects. Oil alternatives, already challenged as they aimed to test new technologies, fuels, and processes, were affected by the same rising costs for feedstocks and construction materials. Production costs fell with the onset of the recession, but some suggest that subsequent economic recovery has already returned industry costs to near previous highs.[32] Rising costs and short-term price fluctuations create uncertainty for those who would invest in new supply, further slowing long term supply growth.

Weather-related and geopolitical disruptions further complicated supply growth. The 2005 hurricanes in the Gulf of Mexico and periodic militant attacks against oil facilities in Nigeria are two examples of such disruptions. Supply disruptions shut down existing supply (sometimes for months or even years), slowed development of new fields, and contributed to concerns about future supply. Figure 5 shows examples of supply disruptions and other events that caused concerns about the adequacy of future supply.

Resource nationalism in oil-rich countries likely limits investment in some easier to produce resources. Relatively easy-to-exploit, inexpensive resources remain in some oil-rich countries that control access to oil exploration and development. Large oil resource holders such as Saudi Arabia, Venezuela, and Russia limit investment from foreign oil companies to varying degrees.[33] According to data reported by EIA, Saudi Arabia, Venezuela, and other members of OPEC hold 72% of the world's oil reserves; Russia, the largest non-OPEC reserve holder, has 4%.[34] The oil industries in these countries are dominated by state-owned oil companies which may manage their resources in ways that result in lower production capacity compared to private sector companies.[35] Also, oil demand within many oil-rich countries is climbing rapidly, reducing how much oil is available for export to other markets. Finally, OPEC countries further limit how much of extant production capacity is utilized.

OPEC's Role in the Oil Market

OPEC countries control their oil output through policy, deciding how much of their oil production capacity to utilize in an attempt to manage oil prices. OPEC country oil ministers meet periodically to set a crude oil production ceiling for the group and allocate production quotas to each of the cartel's members.[36] This leaves them with spare production capacity that can be brought on line relatively quickly.

In contrast, countries outside of OPEC tend to utilize oil production capacity at the maximum level possible because these countries and companies operating in them act as price takers. OPEC is made up of some of the world's largest oil exporters (and largest holders of oil reserves): Algeria, Angola, Ecuador, Iran, Iraq, Kuwait, Libya, Nigeria, Qatar, Saudi Arabia, and Venezuela. The cartel's collective share of global oil production varied between 38% and 42% over the last decade. OPEC members' exports are around 60% of oil traded internationally.[37]

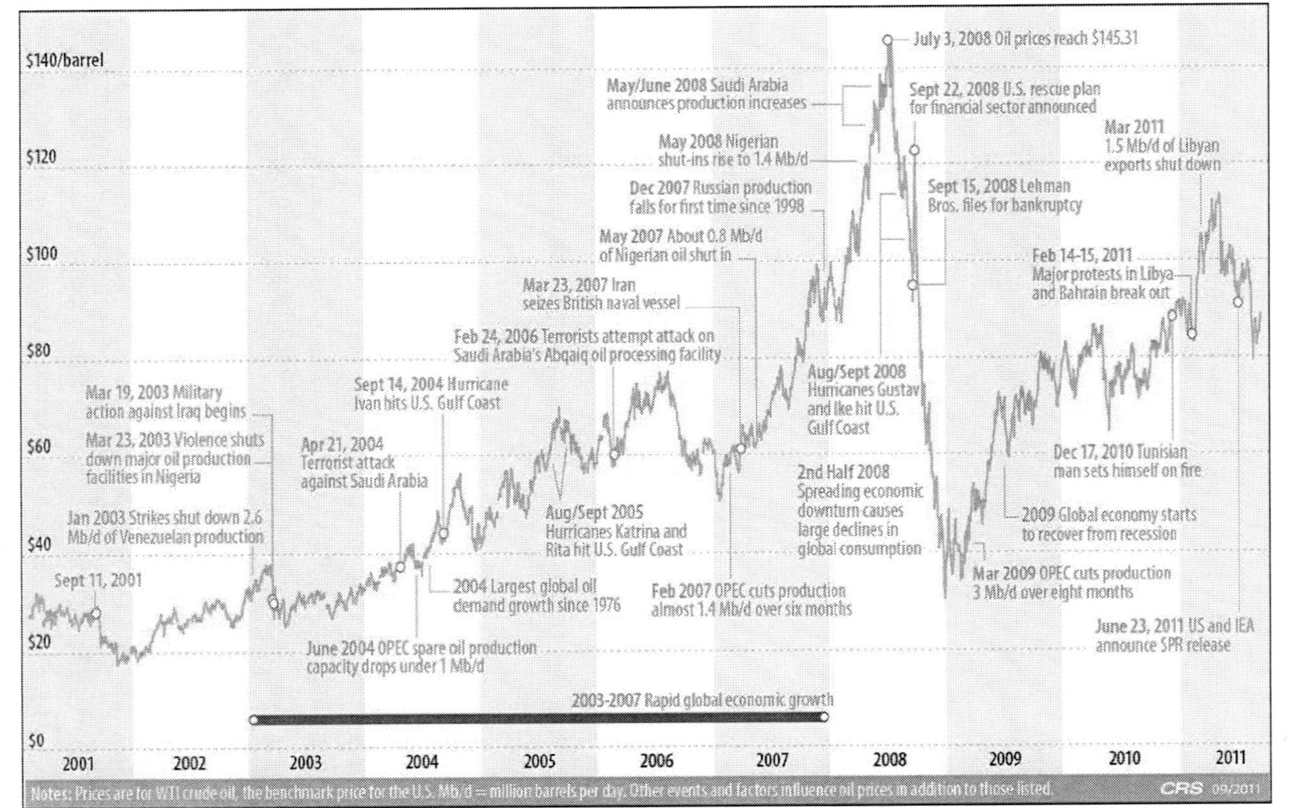

Figure 5. Selected Events and the Price of Oil.

The existence of a cartel in the oil market likely contributes to oil prices being higher on average, but should also contribute to more stable prices, though this is debated. OPEC argues that it contributes to price stability by managing supply to match demand (and holding spare capacity which is discussed below).[38] Critics point to the large price fluctuations over the last decade as evidence that it has not been effective in providing stability. Some have suggested that in the current market environment, OPEC can put a floor on prices, but has limited ability to cap price increases.[39] Others argue that many OPEC members may have limited interest in preventing the upward movement of oil prices and criticize the organization for not producing more oil to moderate prices and for not allowing more of their resources to be developed.

OPEC's willingness to curtail production leaves OPEC countries holding extant but unused crude oil production capacity. OPEC countries—those that obey their quotas and actually maintain this spare capacity—feel the ability that spare capacity gives them to manage prices is worth the cost of developing capacity and leaving it unused. OPEC spare capacity provides a cushion to global markets—supply that can come online quickly in event of unanticipated events such as supply disruptions. However, this is contingent on OPEC willingness to develop and use spare capacity. When OPEC spare capacity declined to less than 1 Mb/d after production ceiling increases in 2004, it contributed to fears about the adequacy of global supply in two ways: First, lower spare capacity increased the risk that OPEC would not be able to offset future supply disruptions. Second, it contributed to concerns that OPEC members could not or would not invest enough in growing production capacity (or allow such investment from private companies) to meet future demand. There were also concerns about the accuracy of capacity estimates and the quality of the crude comprising the available spare capacity—that oil was believed to be relatively heavy, which is more difficult to process.[40]

OPEC spare capacity now sits near 3 Mb/d according to EIA estimates. This is sufficient to offset disruption of all but the world's largest oil exporters.[41] But there are concerns that this may not be enough to also keep up with demand without significant capacity growth in OPEC countries. A key uncertainty remains how Iraq can carry out plans to increase its production capacity.[42]

Market participants frequently focus on the comments and actions of Saudi Arabia. Saudi Arabia is the world's largest oil producer,[43] holds most of the world's spare oil production capacity, and is home to the world's largest conventional oil reserves. With such substantial reserves, Saudi Arabia may take a longer term view of oil markets than some fellow OPEC members and may be more concerned about prices rising so high that it destroys demand in the long term. Just as consuming countries are concerned about security of supply, major oil producers express concerns about security of demand. For instance, Saudi Arabia unilaterally announced it would increase production to moderate price in May 2008, without waiting for OPEC agreement. However, when OPEC announces production cuts, Saudi Arabia frequently ends up accounting for a disproportionately large share of actual reductions in supply.[44]

Peak Oil Concerns

Disappointing supply growth despite dramatic price increases has led to concerns that oil production will soon peak and then decline. Peak oil advocates argue that insufficient new oil is being discovered globally to offset that which is being produced, and that this will result in global production reaching a maximum rate, beyond which output could decline sharply. Some believe this point will occur in the next few years.[45] Peak oil opponents see no peak in oil production resulting from physical resource constraints for the foreseeable future.[46] There is little debate that the world is using oil faster than it is being formed by geochemical processes in the earth's crust (i.e., oil is ultimately a finite resource), and that the oil industry is progressively moving towards more difficult, and often more expensive, resources. Experts disagree on how much oil is ultimately recoverable (including whether this is knowable and how it is affected by technology and investment), how close the world is to peak production, and whether post peak global oil output will decline sharply, fall slowly, or plateau for a period of time before declining.

Rapid increases in oil prices throughout the 20th century have led to periodic predictions that the world had used so much of its recoverable oil resources that production would soon peak and decline regardless of price.[47] These concerns re-emerged in the 2000s. But the world oil supply has continued to increase, with global oil production roughly 8 Mb/d higher in 2010 than in 2000,[48] and it is continuing to rise in 2011. Many forecasters expect oil supply growth will continue for the foreseeable future.[49] However, the sources of oil are changing.

Conventional oil is being discovered in new areas, such as Ghana, the South China Sea, and the Arctic. New technologies make more complex resources commercially accessible, such as hydraulic fracturing to access shale oil deposits in North Dakota and Montana.

Further, oil supply may be increasingly supplemented by use of alternative fuels in transportation such as biofuels, coal or gas converted to liquids, or electricity and natural gas in alternative fuel vehicles; though the abilities of these resources to displace oil do face their own difficulties with scalability.[50] While these various sources of supply appear to be more expensive, they do appear to be coming online to varying degrees, enabled by higher prices.

Note that the oil market could stop growing or shrink for reasons apart from physical depletion of global oil resources, due to "above ground factors." Governments in oil-rich countries could limit exploitation of their oil resources despite physical availability.

Consuming countries' policies on energy security or climate change could raise efficiency, promote alternatives, or raise taxation in a way that could limit demand. The potential for such developments have led some to suggest we could actually see "peak demand."

Some believe that peak demand may already have occurred in the world's advanced economies.[51] However, most analysts see neither peak oil supply nor demand globally at least through 2030, as is discussed in the "Future Fundamentals" section below.

Refining and Crude Oil Prices

Refining constraints can contribute to a premium for light, sweet crudes like WTI as occurred during the mid-2000s.

Rising global demand for petroleum products has been concentrated in middle distillates such as diesel and light distillates like gasoline; consumption of heavy fuel oil has fallen.[52] Demand has also increased for lower sulfur crudes in response to tighter sulfur content standards for fuels in the United States, Europe, and elsewhere.

Lighter, sweeter (less dense, lower sulfur) crudes yield greater proportions of light, low sulfur products and fewer heavier products. But the global supply of crude oil is not getting lighter or sweeter. The average barrel of crude has grown slightly heavier and more sour.[53]

And spare capacity that OPEC can produce is believed to be heavier as well. This trend toward heavy crude created a premium for light sweet crudes like WTI—the U.S. benchmark crude oil—during the oil price run up of the 2000s, widening the differential between the lighter crudes and the heavier, sour crudes.

Complex refining capacity—which can convert heavy, sour crudes into lighter, low sulfur products—was in short supply. Figure 7 illustrates how this premium rose from 2004 to 2008.

Refining Basics

The average yield of petroleum products from a barrel of crude oil refined in the United States in 2010 was 19 gallons of gasoline, 10 gallons of diesel, and 15 gallons of other petroleum products like jet fuel and residual fuel oil (see Figure 6).[54] Yield varies depending on the particular type of crude used and a particular refinery's capabilities. Light crudes are more valuable because they tend to yield more gasoline, diesel and other high value products. In turn they yield less residual fuel oil and other lower value products.

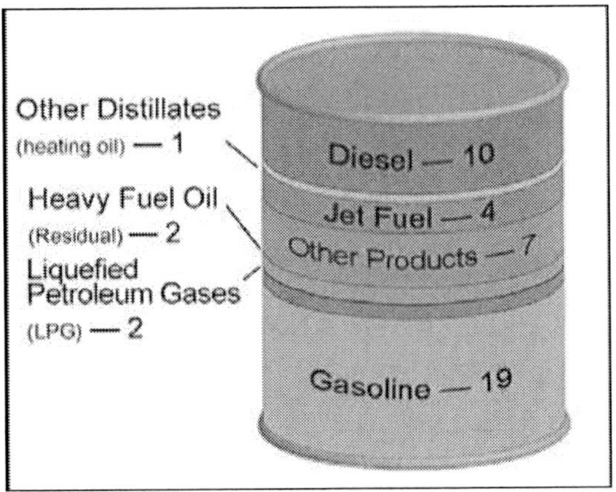

Source: EIA, http://www.eia.gov/energyexplained/index.cfm?page=oil
Notes: Average yield in the United States from refining a 42 gallon barrel of crude oil, which results in slightly more than 42 gallons of refined petroleum products due to volumetric gains when oil is processed.

Figure 6. Average U.S. Product Yield from a Barrel of Crude Oil, 2010 Gallons.

Refiners can invest in complex processing equipment that allows them to further process some lower value products, turning them into more high value products, such as gasoline or diesel. Refiners can also invest in processing equipment that allows them to remove sulfur, allowing them to turn sour (higher sulfur) crudes into products that meet low sulfur fuel requirements that exist in the United States, Europe, and elsewhere. The United States has among the world's most sophisticated refining sectors, one which is geared towards maximizing gasoline yield, as reflected in Figure 6.

> These investments are costly, but allow refiners to use more low-cost heavy, sour crudes. The investment in processing capacity upfront allows lower cost crude acquisition in the future. A refining "margin" is the difference between the cost of acquiring crude oil and the revenue it receives for the mix of products it can produce from the crude oil.
>
> The revenue side of the equation is based the types of fuels a refinery can produce and the price each is selling for in the market. The United States has significant complex refining capacity, particularly in the Gulf Coast region, where refiners have historically received heavy crudes, such as some of the imports from Mexico and Venezuela.

The spread between light and heavy crudes fell when the recession and falling demand loosened oil market conditions. Further, higher refining margins between 2004 and 2008, particularly for complex refineries, led to more investment in refining capacity around the world. The spread between light and heavy crudes has rebounded in 2009, but remains below the levels reached between 2004 and 2008.[55]

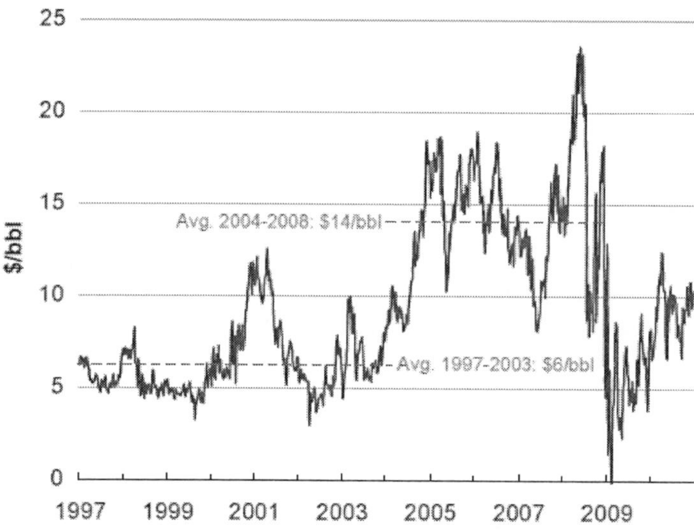

Source: EIA, http://www.eia.gov/dnav/pet/pet_pri_wco_k_w.htm.

Figure 7. Light, Sweet Versus Heavy, Sour Crude Oil Price Spread; The price of WTI minus the price of Mexico's Maya crude oil, a heavy sour crude oil, 1997-2010.

As mentioned above, the crude oil costs account for most of the level and variation of petroleum product prices, including gasoline prices. However, refining and fuel specification issues can further contribute to the cost of petroleum products such as gasoline. For instance, the transition in the United States to ethanol gasoline additives from the additive methyl tertiary butyl ether (MTBE) contributed to higher gasoline prices during 2006.[56] Clean Air Act requirements have prompted 12 states to require 15 unique gasoline blends to accommodate regional and seasonal variations in gasoline combustion. The division of the gasoline market into separate, smaller "boutique fuels" markets can add to fuel costs and contribute to higher, more volatile gasoline prices at times. For example, a market disruption can have larger effects in a segmented market than a single, larger market.[57] These requirements were put in place to improve public health through reducing the impact of vehicle emissions on air quality.

WTI Price Depressed

WTI is the main U.S. benchmark crude and as such is the crude price referenced in this report, but it is worth noting that WTI price has been depressed relative to other oil prices due to recent developments in the central United States. Due to its better quality, WTI has historically traded at about a $1.50/bbl premium relative to Brent crude, the main European benchmark crude price. [58] But since August 2010, this spread has inverted. Brent has averaged about $13/bbl higher than WTI in the first half of 2011. Rising oil supply in the Midwest—from the Bakken Formation and Canadian oil sands—and lower U.S. demand has created a glut of oil at Cushing, Oklahoma, the hub at which WTI is priced. There is insufficient pipeline capacity to carry this oil to the Gulf Coast region, where it would be reconnected with international markets. Much of the recent WTI discount reflects the cost of transporting crude from the Midwest to the Gulf Coast by truck or rail, a relatively expensive means of transport.[59]

The price differential offers a financial incentive to build more pipeline capacity from the Midwest flowing toward the Gulf Coast or reverse existing pipelines currently flowing the opposite direction. But according to reports, individual producers may be reluctant to commit their own capital to such a project in hopes that that someone else does so.[60] Some with existing pipeline capacity flowing from the Gulf Coast up to Cushing are reluctant to reverse their pipelines.[61]

> One proposal which may alleviate some of the glut at Cushing is the Keystone XL pipeline, which would carry crude oil from Canadian oil sands through Cushing and on to the Texas Gulf Coast. The Keystone XL project awaits approval from the U.S. Department of State, which issued a final Environmental Impact Statement on the pipeline in August 2011 and aims to make a final determination on if the pipeline is in the U.S. national interest by the end of the year.[62] Keystone XL has been opposed by environmental concerns regarding oil spill risks and general opposition to Canadian oil sands development. For more on Keystone XL, see CRS Report R41668, *Keystone XL Pipeline Project: Key Issues*, by Paul W. Parfomak et al.

Prices Collapse with Recession, Rebound with Recovery

The global economic downturn started in the U.S. recession in late 2007, but U.S. GDP did not start falling outright until the second quarter of 2008. When it did start falling, the decline was rapid and contagious—by the third quarter of 2008, global GDP was contracting.[63] The financial crisis reached crescendo at this time with Lehman Brothers filing for bankruptcy in September 2008. Oil demand fell with economic activity. Meanwhile, oil supply was increasing. Several OPEC countries had increased production during the price run up. By May of 2008, Saudi Arabia expressed concerns about how high prices had gone and unilaterally announced that it would increase crude oil supply.

Prices fell from a peak of over $140/bbl in early July, 2008 to almost $30/bbl by year end. (See price fluctuations in Figure 5.) Non-OPEC supply is price inelastic, especially in the short run, and even OPEC cuts can take months to be fully implemented and reach markets, leaving few forces to balance the market except a sharp decline in prices. Global oil consumption stopped deteriorating by the second quarter of 2009, which is also when OPEC cuts, announced in late 2008, were fully implemented and reaching consuming markets. Still, demand growth was moderate and production cuts had left OPEC with ample spare capacity. This helped keep prices between $70 and $80/bbl for most of the period from July 2009 to October 2010.

By the fourth quarter of 2010, global consumption surpassed pre-recession levels, again driven by the emerging markets. Oil prices responded by firmly breaking past the $80/bbl level in late 2010. They jumped up further in early 2011 after pro-democracy movements led to civil unrest spreading from

Tunisia to Egypt and then to Libya. Libya's 1.5 Mb/d of exports were shut-in and concerns arose that the "Arab Spring" would disrupt exports from other key producers in the region.[64] Rising demand and new risks to supply again pushed prices up to near $100/bbl and higher between March and July 2011. Prices in August moved lower with renewed economic weakness.

June 2011 Strategic Petroleum Reserve Release

The Libya disruption in February 2011 led to calls by some legislators for President Obama to release oil from the Strategic Petroleum Reserve (SPR).[65] The President is authorized to use the SPR in event of a "severe energy supply interruption or by obligations of the United States under the international energy program of the International Energy Agency (IEA)."[66] The Administration initially held off on releasing crude from the SPR, but promised readiness to release oil if conditions worsened,[67] eventually announcing a release of 30 million barrels from the SPR on June 23 in coordination with strategic stock releases of about 30 million barrels from other members of the International Energy Agency (IEA) to be released over 30 days.[68]

The IEA release of oil from reserves aims to offset some of the market effects of Libya's prolonged supply outage. The decision followed the June 8, 2011 OPEC meeting, where OPEC members failed to agree on whether to increase their production ceiling. The IEA may have wanted to make clear to OPEC that they felt additional supplies were needed. In their announcement, however, the IEA made clear it appreciated increased output from producers who would increase supply and suggested its release will help bridge the gap until these resources reach global markets.[69]

Given the range of factors constantly affecting oil prices, it is difficult to isolate the price impact of the IEA action. Oil prices dropped about $5/bbl immediately following the announcement but climbed back to pre-release levels by the end of June; though they may have been lower than would otherwise have been without the release. Prices did fall more dramatically in early August, soon after the release was completed, but the dominant driver was likely worsening economic expectations.

Some have argued that the release should have come sooner,[70] while others suggest that the release was not justified.[71] This is the third time that the United States and other IEA members have made a coordinated release of strategic stockpiles.[72]

Some IEA members were considering a second release of strategic reserves, though others were opposed.[73] European refiners may not favor such releases because, unlike the United States, most European strategic stocks are held as refined products (the release of which could dampen refiner profits). With prices dragged down by economic events in August, 2011 the issue may be moot.

Future Fundamentals and the Price of Oil

There are many variables involved with how long-term supply and demand developments will contribute to future oil price trends. Shorter term price variation around any long-term trend is likely to continue.

EIA's reference case oil market projections, which assume existing policies only, have WTI rising to $125/bbl by 2035 in terms of 2009 dollars—$199/bbl in nominal terms. However, EIA underscores the uncertainty of oil prices by providing high and low price scenarios, which differ in economic and resource assumptions, that have prices rising to $200/bbl and falling to $50/bbl by 2035 in 2009 dollars, respectively. Many oil market forecasters see oil prices in the future higher in nominal terms; however, opinions differ with regard to inflation adjusted prices based on expectations about future fundamentals. Developing countries are likely to continue to consume more oil in order to fuel their economies, and political constraints in resource rich countries are likely to persist. These trends would contribute to higher future oil prices. However, their impacts can be offset by new policies, greater investment, or technology developments that expand production from new resources, reduce the cost or expand the scale of alternative or unconventional supplies, or enable greater efficiency improvements. Such developments in the United States can reduce risks to the economy and energy security by reducing the need for U.S. imports, and can contribute to lower oil prices by increasing supply and reducing demand in the global oil market where prices are set. Efficiency improvements have the additional benefit of insulating consumers by reducing the need for oil or other fuels even if their prices are driven up by other market events. Some of this is already occurring and has contributed to lower U.S. oil imports in recent years. For instance, technology improvements are enabling greater domestic oil production and new vehicle fuel efficiency is climbing ahead of regulatory requirements.[74] Key U.S. oil-related supply and demand policy measures currently being considered are discussed below.

Whatever the long-term trajectory of prices, short term price variations, both up and down, around that trend will continue. Few anticipate an end to the economic cycles, supply disruptions, and other developments that have set off price movements over the last decade. The short-term price inelasticity of supply and demand can lead such events to prompt wide swings in oil prices. Such swings could contribute to higher longer term oil prices if the uncertainty they create undermines the commitment of companies and governments to long term investments in new supply or greater efficiency.

Price inelasticity could also potentially allow for other factors, not related to the physical supply and demand, to raise (or lower) oil prices. Slow, price-inelastic supply and demand adjustments imply that market fundamentals may not react quickly to correct pricing aberrations. Non-fundamental factors that may cause pricing aberrations have been of great concern to Congress in recent years, especially where they may involve market participants trying to unfairly profit from already high oil prices. These factors are discussed in the next part of this report, "Non-Fundamental Factors: Speculation, Manipulation, and the Dollar."

Supply and Demand Policy Measures

More than 100 oil-related bills have been introduced in the 112th Congress. Several bills aimed at affecting domestic supply (conventional and alternative) and demand have seen major legislative action, many of which focus on expanding potential offshore oil and natural gas development with the intention of reducing oil imports and contributing to lower oil prices through greater domestic oil production.

Several oil-focused bills have been passed by the House of Representatives.[75] H.R. 1230, the Restarting American Offshore Leasing Now Act, would require the Secretary of the Interior to conduct four lease sales within about a year of the bill's enactment.[76] A second bill, H.R. 1231, would require lease sales in the most prospective areas in each of the Outer Continental Shelf (OCS) Planning Areas for the 2012-2017 5-Year Leasing Program, specifically, areas expected to contain more than 2.5 billion barrels of oil or more than 7.5 trillion cubic feet of natural gas. H.R. 1229 would provide a new safety review and seek to expedite the drill permitting process by providing a new timeline for the Secretary to make a final decision on the permit application. The bill includes language on judicial reviews that would provide timelines, an exclusive venue for civil actions, and limits on relief and attorney fees. H.R. 2021, the Jobs and Energy Permitting Act of 2011, amends the Clean Air Act to limit the scope of air quality impact measures of offshore developments. Separately, H.R. 910, which limits the scope of EPA action on greenhouse gas regulation, includes provisions to preserve vehicle efficiency requirements (which the administration had tied to EPA greenhouse gas regulation, see *Demand Efficiency* below).

Two oil-focused bills reached the Senate floor for debate but failed to achieve 60 votes for cloture. S. 953, the Offshore Production and Safety Act of 2011, was similar to H.R. 1229. Separately, S. 940 would remove certain tax benefits for large oil and gas companies, and repeal the Secretary of the Interior's authority to suspend royalties for certain kinds of oil and gas drilling.

Two additional bills requesting studies have been reported to the Senate from the Energy and Natural Resources Committee. S. 1343 requests a study by the National Academy of Sciences on the water used in production of oil and other resources, including domestically produced oil, imported oil, shale oil such as that from the Bakken Formation, biofuels, and coal-to-liquids supply. S. 916 directs the Interior Department to conduct a comprehensive inventory of U.S. oil and natural gas resources, including Atlantic, eastern Gulf of Mexico, and Alaska regions of the outer continental shelf.

Many additional oil-focused bills on a range of subjects have been introduced but have yet to see major action. Key issues include:

Domestic Oil and Gas Drilling

The OCS bills above are only a few of the bills introduced in the 112th Congress which aim to increase domestic oil production. A number of other such bills focus on opening the Arctic National Wildlife Refuge (ANWR) to oil and gas drilling. Proponents of opening ANWR or more of the OCS argue that increased domestic production of oil would reduce imports, may reduce the price of oil by increasing supply, and could increase tax revenues and create jobs in the energy industry. Opponents point to environmental risks, claim any price impacts would be limited,[77] and suggest that alternatives to oil, such as electric vehicles, should be prioritized as a long term solution that may also create jobs. Seven bills have been introduced concerning ANWR in the 112th Congress. Five bills (H.R. 49, S. 351, S. 352, H.R. 909, and H.R. 1023) would open the area in whole or in part to development, and two (H.R. 139 and S. 33) would designate the area as wilderness.

Separately, there is debate around the use of hydraulic fracturing, which has contributed to recent increases in the production of oil and other liquid hydrocarbons. Hydraulic fracturing involves injecting water, sand and chemicals into low-permeability geologic formations to fracture the rock and allow oil and natural gas to flow to the well. There is concern about environmental effects of chemicals used in hydraulic fracturing on water supplies. While the hydraulic fracturing debate is frequently focused on natural gas supply, high prices for liquid fuels has kept the oil and gas industry focused on natural gas investments that also produce natural gas liquids (NGLs). NGLs supplement domestic oil supply. Hydraulic fracturing has also been used to stimulate production of shale oil, particularly in the Bakken Formation underlying parts of North Dakota and Montana. At present, it is regulated at the state level. The Energy Policy Act of 2005 exempted hydraulic fracturing from Safe Drinking Water Act (SDWA). H.R. 1084 and S. 587 would repeal the exemption for hydraulic fracturing, thus authorizing EPA to regulate this process under the SDWA. The bills also would require disclosure of the chemicals used in the fracturing process.

Demand Efficiency

About 72% of the oil consumed in the United States is used in the transportation sector, mostly as gasoline. Consequently, efforts to improve oil efficiency focus on the use of oil in transportation. An important development was the increase of the Corporate Average Fuel Economy (CAFE) Standards in the Energy Independence and Security Act of 2007 (EISA). CAFE standards set efficiency requirements for new vehicles. EISA authorized the first increase in CAFE standards for more than two decades; it included provisions to increase the CAFE standards for light duty vehicles (LDVs) to 35 miles per gallon (MPG) by 2020. The Obama Administration accelerated the CAFE target by four years, to 2016, and integrated CAFE targets with automotive greenhouse gas emissions standard issued by the Environmental Protection Agency (EPA). On July 29, 2011, EPA and the National Highway Traffic Safety Administration (NHTSA) announced its intention to raise CAFE standards for LDVs for the period from 2017 to 2015, eventually reaching to 54.5 MPG.[78] The plan raises CAFE standards about 5% a year between 2017 and 2025. An earlier EIA analysis of CAFE standards examined a slightly more aggressive plan, raising the standard at 6% a year from 2017 to 2025, could reduce expected 2035 oil consumption by about 3 Mb/d, oil imports by about 2 Mb/d, and oil prices by $6/b versus their baseline forecast.

The administration also announced the first ever fuel standard for heavy-duty vehicles. Efficiency requirements vary by class of vehicle, rising 10-20% between model years 2014 and 2018. Heavy-duty vehicles like tractor trailers and other delivery vehicles account for 12% of U.S. fuel use.[79]

Several bills have been introduced in the 112th Congress which propose to further improve demand efficiency, including bills that provide new tax credits for fuel efficient vehicles,[80] allow state and local governments to prescribe fuel economy for taxis,[81] and require a federal plan to reduce the government's petroleum consumption.[82] Of note is that bills in the 112th which aim to prevent the EPA from pursuing greenhouse gas emissions controls include provisions which preserve CAFE targets under the authority of the Department of Transportation.[83]

Certain policies may also provide incentives for efficiency through raising the cost of fuel consumption, thereby providing consumers a price signal to improve efficiency. Proponents of raising federal gasoline and diesel taxes cite both efficiency benefits and the need to increase funding for the Highway Trust Fund for which these funds are earmarked.[84] However, there are concerns that any increase in the current federal fuel taxes could be unpopular or, depending on how they are structured, may negatively impact the economy. The current federal excise tax on gasoline and diesel is set to expire on September 30, 2011.

Alternative Fuels

Much of the attention to alternative fuels in the 112th Congress has focused on biofuels, particularly on whether to extend the 45 cent per gallon Volumetric Ethanol Excise Tax Credit (VEETC) available to gasoline suppliers for blending ethanol. Concerns include VEETC's high fiscal costs,[85] that the ethanol industry is a relatively mature industry and may not need VEETC, and that ethanol volumes have already reached the maximum levels that can be blended into the gasoline pool.[86] VEETC proponents argue domestic ethanol's energy security benefits; it has been a significant part of domestic liquid fuels supply growth in recent years. The tax credit is set to expire at the end of 2011; some legislative measures aim to end it sooner.[87] Measures have also been introduced to support other alternative transportation fuels, including electricity,[88] natural gas,[89] and coal.[90]

Strategic Petroleum Reserve

Prior to the emergency release of the SPR in June 2011, the President's 2012 budget proposal had included a $500 million, approximately 6 million barrel, non-emergency sale related to damage at one of the caverns that comprise the SPR.[91]

Several legislative proposals in the 112th Congress aim to change the composition of the SPR in attempt to better meet U.S. energy security needs. Provisions in H.R. 1748 (taken from H.R. 1017, the Enhanced SPR Act) would require the substitution of some crude oil in the Strategic Petroleum

Reserve (SPR) with refined products reserve. H.R. 142, the National Strategic Gasoline Reserve for Purposes of National Security Act of 2011, would create a national gasoline reserve. H.R. 1914, the Gas Price Stabilization Act of 2011, and provisions of H.R. 1861, the Infrastructure Jobs and Energy Independence Act, would replace light crude oil in the SPR with heavy crude oil to create a mix of crude oils that better reflect the qualities of crudes refined in the United States.

Foreign Policy

With so many international developments contributing to higher oil prices, foreign policy can play an important role in addressing oil price risks. However, it can be difficult to reach agreement with other countries about specific response measures. The United States aims to influence international decisions related the oil market through bilateral and multilateral engagement of major oil producers and consumers.

The International Energy Agency (IEA) is a key part U.S. oil-focused foreign policy. In response to the 1973 oil crisis, members of the Organization for Economic Co-operation and Development (OECD) created the IEA. The IEA coordinates its members' responses to oil market emergencies. Members are required to hold strategic stockpiles of oil and petroleum products to be released when oil market emergencies occur. To date, the IEA has released strategic stocks three times: In 1991 during the First Gulf War, in 2005 after Hurricanes Katrina and Rita struck the U.S. Gulf of Mexico, and in 2011 after civil unrest in Libya led to a prolong outage of oil exports. The IEA also provides energy statistics and forecasts which regularly address the amount of oil needed from OPEC to balance supply and demand.

The International Energy Forum (IEF) seeks to create a venue for dialogue between members of IEA, OPEC, and other oil exporting and importing nations. The United States is one of its 86 members. Its flagship program has been the Joint Organizations Data Initiative (JODI) which aims to improve oil market transparency through collecting and publishing information on oil production and consumption.

Improving international oil market information can improve price formation. Much of the international oil market data published by the IEA, IEF, EIA, and others comes from governments. OECD governments publish extensive oil market data in a relatively timely manner.[92] However, data for key developing country producers and consumers is more limited. The IEA and the United States have called for greater transparency from countries such as China.[93] According to a recent IEF report, "Data transparency is essential to efficient oil and gas market stability. Greater transparency aids in price discovery and limits volatility, thus reducing uncertainty for investors."[94] Greater, more timely data on oil production, consumption, trade and especially oil inventories for developing countries is likely to help make sure that oil price fluctuations more accurately reflect supply and demand realities.

The IEA, the United States, and others have also called for countries to remove policies that can distort markets in a way that contributes to higher oil prices. Such policies are not limited to resource development restrictions by oil rich countries. There has been greater attention to energy consumption subsidies, which can reduce the incentive for consumers to be more efficient.[95]

Bilateral relations with key oil producers and consumers, such as Saudi Arabia or China, often include discussions on oil markets factors. However, this may be one among many U.S. policy priorities stressed in a bilateral relationship. Other U.S. foreign policy priorities may include economic, security, or human rights concerns in these countries. Several bilateral energy programs have been created to advance shared goals with international partners. Programs with China have been of particular interest due to China's rapid growth in oil consumption and shared concerns about energy security. Such programs with China include the U.S.-China Oil and Gas Industry Forum and the U.S.-China Clean Energy Research Center (CERC). The U.S.-China Oil and Gas Industry Forum, started in 1998, brings together government and industry officials from both countries to discuss development of secure, reliable and economic sources of oil and natural gas while facilitating investment in the energy industry.[96] CERC aims to invest $150 million from public and private sources in both countries to advance technology in several areas, including technologies with potential to reduce the dependence of vehicles on oil and improve vehicle fuel efficiency.[97]

NON-FUNDAMENTAL FACTORS: SPECULATION, MANIPULATION, AND THE DOLLAR

The rising price of oil has also raised concerns that issues apart from supply and demand—nonfundamental factors—increased prices even higher than what may be justified by supply and demand. The inelasticity of supply and demand may enhance the likelihood that other factors could contribute to higher prices. Of particular concern have been manipulation of physical prices, manipulation of financial oil derivative contracts, the general impact of financial speculation, and the depreciation of the dollar. Concern has led to several bills being introduced in the 112th Congress and to certain Administration actions.

Different federal agencies have been charged with taking the lead on monitoring the non-fundamental factors. The U.S. Energy Information Administration tracks market developments, explains price movements, and provides oil market forecasts. The Federal Trade Commission (FTC) monitors the action of commercial participants in the physical oil market and pursues civil violations. The Department of Justice pursues potential criminal violations. The Commodity Futures Exchange Commission (CFTC) monitors financial contracts related to oil.

On April 21, 2011, the Obama Administration announced creation of a multi-agency Oil and Gas Price Fraud Working Group within the Financial Fraud Enforcement Task Force.[98] The Department of Justice-led group also includes representatives from National Association of Attorneys General, CFTC, FTC, the Department of the Treasury, the Federal Reserve Board, the Securities and Exchange Commission, as well as the Departments of Agriculture and Energy. The group aims to explore whether there is any evidence of manipulation of oil and gas prices, collusion, fraud, or misrepresentations at the retail or wholesale levels that would violate state or federal laws and that has harmed consumers or the federal government as a purchaser of oil and gas. It will also evaluate developments in commodities markets including an examination of investor practices, supply and demand factors, and the role of speculators and index traders in oil futures markets.

The Role of Speculators

As oil prices were rising in the 2000s, the markets for financial contracts linked to the price of oil also expanded. Volatile oil prices attracted speculative investments from a wide variety of banks, asset managers, pension funds, endowments, hedge funds, and other financial institutions. Financial speculators trade oil derivative contracts among themselves and with hedgers, or companies whose primary business involves the production, processing, transportation, or use of oil and petroleum products.[99] Hedgers use or produce physical oil. They use financial contracts to lock in today's oil price for future transactions, thus protecting their profitability from oil price fluctuations. Speculators generally do not deal in physical oil, but use energy derivatives to take bets on oil prices moving in a certain direction, or in some cases to achieve portfolio diversification.

Historically, volatility in commodity prices has bred suspicion of speculators, who impose another level of economic decision-making between producer and consumer, and who may profit from high prices that hurt consumers and the economy. Nonetheless, two benefits that separate speculation from mere gambling are widely recognized: First, speculators can provide liquidity to hedgers. That is, they are willing to take on the risks that hedgers wish to avoid, providing businesses with protection against price shocks. Second, speculators can contribute to the price discovery process. Their trades can reflect additional information and analysis about market conditions that in theory could make price-setting more efficient.

Derivatives prices and physical prices for oil are linked for several reasons: oil companies and other physical market participants base their pricing decisions on the price of financial oil contracts, and there is the option to take physical delivery on some crude oil contracts, which creates arbitrage opportunities if there is a deviation between physical and futures prices.[100] When U.S. headlines announce a sharp jump (or drop) in the price of oil, the press is frequently referring to the price of a futures contract based on West Texas Intermediate crude oil, traded on the New York Mercantile Exchange (NYMEX). Thus, the expectations of financial speculators trading derivatives are an important factor in determining the prices that consumers pay for oil-related products.

Many observers blame price spikes in 2008 and early 2011 on the influence of speculators. Figure 8 below shows that speculative traders ("non-commercials") have increased their share of all NYMEX crude oil contracts outstanding (the "open interest") over the past decade.[102] The argument is made that a market dominated by financial speculators, rather than firms actually involved in the physical oil business, is likely to misprice oil because speculators may trade on factors unrelated to fundamental supply and demand conditions.

The impact of speculation on commodity prices is an old question in economics, the subject of a large body of academic literature, but not yet a settled point. The bulk of empirical evidence suggests that the presence of speculators or the existence of a futures market does not tend to make prices more volatile.[103] At the same time, the existence of speculative bubbles, when prices can diverge from fundamental values for months or years, is well known. While the rise and fall of oil prices in 2008 can be analyzed as a response to supply and demand factors, it is also possible to consider that episode as a speculative bubble, during the course of which all fundamental news became secondary to the fact that prices had been rising and might rise further.

Financial Derivatives in Commodities Markets

Derivative contracts are so called because their value is based on the price of some underlying commodity. There are several kinds of contracts. An oil futures contract allows an investor to buy (or sell) oil at today's price for delivery at a future date. If the price of oil rises in the interim, the buyer makes money on the contract (and the seller loses money).

In practice, nearly all futures are settled in cash—the parties to the contract close out their positions before the contract expires rather than take physically delivery of the oil.

Through a futures contract, a company which uses oil in its business can insure against future price increase of oil—if the price of oil rises, it will cost the company more to buy oil for its business, but the money it makes on the futures contract can offset the extra costs.

A financial institution can also buy a futures contract, but without any commercial exposure to the price of oil, it may simple make or lose money depending on how the oil price changes.

An option contract differs from a futures contract in that requires payment now for the right to buy an underlying asset at a given price during a certain time period. For example, a company which uses oil for its business may buy an option that essentially gives it the right to buy oil at $100/bbl.

If the price of oil rises, say to $105/bbl, the company has the right to exercise the option and receive the oil, which it can then resell for a $5/bbl profit. (In practice, options contracts are also financially settled, with exchange of the $5/bbl between counterparties rather than the exchange of the underlying asset.)

Futures and options are standardized contracts traded in financial exchanges in New York, London, Singapore, Dubai, and elsewhere. There are other derivatives contracts, swaps and forwards, which are bought and sold over-the-counter as tailor made contracts between counterparties.[101]

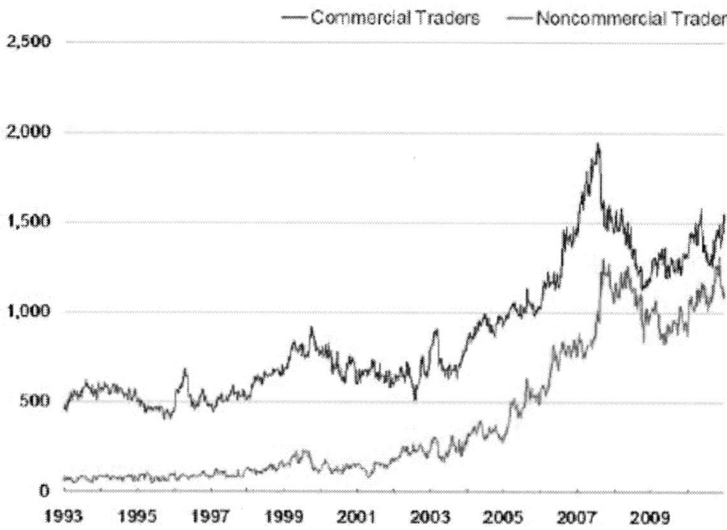

Source: CFTC, Commitments of Traders reports.
Notes: Figures include the sum of long, short, and spread positions.

Figure 8. Futures Contracts Held by Hedgers and Speculators, 1993-2010; NYMEX WTI Oil Futures Contracts Outstanding, Held by Commercial Traders (Hedgers) and Non-Commercial Traders (Speculators), 1000s of Contracts.

Detailed empirical analysis of this and other price episodes may not settle the debate over speculation; data are subject to conflicting interpretations. Changes in futures positions by some financial investors are clearly correlated with the price of oil, but this does not prove causation— that financial investors are driving prices higher.[104] Some suggest there is causation—that capital inflows from financial investors can bid up the price of oil.[105] Opponents argue that there is no empirical evidence of financial investors causing an increase in oil prices; that they may just be following oil prices which are rising for fundamental reasons.[106] Still others argue that, while fundamentals are the main driver of oil prices, financial investment can lead to prices overshooting levels justified by fundamentals for short periods of time, such as in 2008.[107]

Regulation of Speculators

The responsibility for monitoring energy derivatives markets falls primarily with the CFTC. It can sanction market participants found to have manipulated oil prices, through knowing action to create "artificial" prices or by providing false information that might affect prices. The CFTC has brought a small number of manipulation cases in energy markets in recent years. For example, in May 2011, the CFTC announced charges against trading firms Parnon, Arcadia Petroleum, and Arcadia Suisse[108] alleging that these firms manipulated prices by taking a large futures position and simultaneously purchasing a large part of the supply of physical oil that could be delivered to satisfy those contracts.[109] The CFTC charges that the firms made $50 million in illegal profits by artificially increasing the price differential between two futures contracts by nearly $2/bbl for several days during January and March of 2008.

Even if proven, this case involves a short term price impact. This is typical of manipulation cases: In the wake of the Enron scandal; several energy firms were charged with providing false information to data providers whose reports are followed by traders. For example, a report that understated the amount of natural gas in storage might cause prices to tick upwards, but only until the next report was issued. Short term price dislocations may be very profitable to the manipulator, but they do not explain price trends that last for months or years. The CFTC has never found any evidence that major price spikes such as that in 2008 were the result of deliberate manipulation by financial firms.

Federal commodities law states that "excessive speculation" causing "sudden or unreasonable fluctuations" in commodity prices is "an undue and unnecessary burden on interstate commerce."[110] While excessive speculation, unlike manipulation (discussed below), is not a violation of law per se, the CFTC is given tools to prevent it from occurring. The principal such tool is position limits, which are intended to prevent any speculator from amassing a position large enough to obtain market power or the ability to influence prices. Position limits in energy contracts are set and administered by the exchanges, although the Dodd-Frank act, when implemented requires the CFTC to set limits itself for all physical commodities.[111] In addition to position limits, CFTC receives daily reports on large positions in the markets, allowing it to monitor suspicious trading patterns.[112]

The CFTC's authority to control manipulation and excessive speculation is generally structured to prevent a single large trader from dictating prices. If the contention is that a number of speculators, each acting within the law, have a collective impact that distorts prices, then the CFTC's possible response is less clear. It could (as opponents of oil speculation urge) tighten position limits significantly, forcing large speculators to liquidate parts of their positions, but has not taken such action in the past.[113] Until the agency finds evidence that speculation is causing price distortion, it is unlikely to take actions that could reduce liquidity or otherwise disrupt markets. Recent legislation, however, gives the CFTC more authority to control speculation and manipulation.

Recent Legislation

The Dodd-Frank Wall Street Reform and Consumer Protection Act of 2010 (P.L. 111-203) includes measures to increase limitations on speculation, improve market transparency, raise penalties on market manipulation, and give financial regulators more authority to address speculation.[114]

- *Margin.* The CFTC is given limited authority to set margin levels on the futures exchanges. (Margin is the amount of cash that traders must deposit with their brokers before opening a position; higher margins raise trading costs.) Previously, CFTC could change margins only in emergencies. (Section 736)
- *Position Limits.* The CFTC is directed to establish position limits for both swaps and futures. (CFTC has long had authority to set limits on the size of futures positions, but has generally delegated this function to the exchanges.) Position limits may now apply to groups or classes of traders (as well as individuals). (Section 737)
- *Anti-manipulation Authority.* New prohibitions against manipulation by means of false reporting or false information. (Section 753)
- *Foreign Boards of Trade.* Foreign futures exchanges offering direct electronic access to their trading systems to U.S. persons must maintain rules regarding manipulation and excessive speculation comparable to those in U.S. law and regulation and must provide the CFTC with full market information. (Section 738)

Beyond these specific provisions, Dodd-Frank is to bring increased transparency to previously-unregulated over-the-counter swap markets. CFTC is to receive full information on all commodity swap transactions, including those that are not traded on organized markets, and thus be able to monitor the derivatives market in a more comprehensive fashion, since swaps, futures, and options are used interchangeably by traders and may all have an impact on price. The CFTC is in the process of developing regulations and rules authorized by Dodd-Frank, but full implementation will not occur until 2012, at the earliest.

Two approaches to address potential price effects of financial speculation in the 112th Congress are illustrated in H.R. 2003 and S. 1200/H.R. 2328. H.R. 2003, the Taxing Speculators out of the Oil Market Act, discourages speculation by raising the cost of derivatives trading by financial institutions with a one-cent tax on their derivatives transactions. S. 1200 and H.R. 2328, both titled the End Excessive Oil Speculation Now Act of 2011, aim to accelerate implementation of Dodd-Frank limits on speculation in oil futures markets. The bills would require the CFTC to impose position limits and to raise margin levels for speculators within 14 days, regulations which would remain in place until position limits and margin requirements authorized under the Dodd-Frank Wall Street Reform Act are established. It authorizes (but does not require) the CFTC to use its position limit authority to reduce speculative positions to less than 35% of all positions in energy derivatives.

Manipulation of Physical Prices

There is public concern that suppliers may manipulate petroleum supply or otherwise unfairly raise the price of oil and oil products. Among the ways a company could attempt to do this is to become large enough in a particular market to prevent competitive pricing, collude with other companies to similar effect, report false information that may move prices in their favor, or take advantage of emergency situations to raise prices more than warranted.

"Price gouging," or inflating prices to "unfair" levels in order to take advantage of certain circumstances causing a decrease in supply, has been of particular interest to Congress and the public since 2005, after gasoline prices rose in the wake of hurricanes in the southeastern United States.[115] Currently, no federal law specifically addresses price gouging. However, at least 30 states have price gouging laws on the books.[116] Most states have laws that are triggered in the event of a declared emergency, with a few having laws that may be applicable at other times as well. Generally, these laws prohibit the sale of goods and services in the designated emergency area at prices that exceed the prices ordinarily charged for comparable goods or services in the same market area at or immediately before the declaration of an emergency. However, many statutes include an exemption if the increased prices are the result of increased costs incurred for procuring the goods or services in question.

While there is currently no federal anti-price gouging law, antitrust and anti-competitiveness laws may apply to unfair manipulation of physical oil prices.[117] The FTC is responsible for identifying, prosecuting, and preventing any unlawful anticompetitive practices or mergers within the oil industry.[118] It analyzes proposed mergers and challenges transactions that likely would substantially reduce competition or result in higher prices of crude oil or petroleum products.[119]

The FTC also polices anticompetitive conduct in petroleum markets that includes monitoring price movements and other activity to identify any that may not reflect the normal workings of competition. The agency actively monitors fuel price movements in 20 wholesale regions and approximately 360 retail areas across the country. The FTC initiates law enforcement investigations in response to suspect pricing episodes as they are identified. In response to requests from Congress and President George W. Bush, the FTC has conducted two special investigations into gasoline price manipulation since 2005. On both occasions it found no instances of illegal market manipulation causing higher prices, instead attributing price increases to supply and demand related developments.[120] To limit the risk of manipulation in the future through false reporting, Congress authorized the FTC to prohibit potential manipulative or deceptive conduct in wholesale petroleum markets through misleading statements by energy industry participants in the Energy Independence and Security Act of 2007.[121] The FTC's subsequent Petroleum Market Manipulation Rule prohibits fraudulent or deceptive conduct that could harm wholesale petroleum markets.[122] It was published in August 2009 and has not yet led to any enforcement actions.

Several measures to further address potential physical price manipulation have been introduced in the 112th Congress. H.R. 1899, Oil Consumer Protection Act of 2011, amends the Clayton Act to make unlawful the export, diversion, or otherwise withholding of energy supplies from the U.S. market with the intention of raising domestic energy prices. H.R. 1899 also incorporates the Federal Price Gouging Prevention Act (H.R. 964 and H.R. 1748) to make unlawful any excessive increases in oil prices during a crisis in the oil market.[123]

H.R. 1899, along with S. 394, the No Oil Producing and Exporting Cartels Act, contains provisions to apply the Sherman Antitrust Act to OPEC. Separately, H.R. 1347, the Oil Price Reduction Act of 2011, prohibits U.S. bilateral assistance and arms exports to any country that engages in oil price fixing to the detriment of the U.S. economy.

Similar "NOPEC" legislation has been introduced in several previous Congresses. Such measures have faced opposition due to concerns about its effectiveness, risk of retaliatory measures, and potential impacts on U.S. diplomatic interests, U.S. investments abroad, and foreign investment in the United States.[124]

The Dollar and the Price of Oil

The relationship between the U.S. dollar and oil prices is complex. Some analysts argue that the declining value of the U.S. dollar causes an increase in the price of crude oil. This is a regular refrain in media reporting on oil prices and financial markets. But the relationship is likely not as important, and certainly not as straight forward, as news reports sometimes suggest.

Oil is largely priced in U.S. dollars. When the U.S. dollar depreciates, consumers in foreign countries with appreciating currencies do not see the same price increase as consumers in the United States. These foreign consumers presumably go on to consume more oil than might otherwise have been the case if they experienced the same price increase as Americans. In turn, this higher consumption could drive up the price of oil. But some of the largest oil consumption increases have actually come from countries that index their currency to the dollar to varying degrees, such as China and Saudi Arabia. Consumers using the Euro—which appreciated versus the dollar during some periods when oil prices were rising—reduced their consumption along with other advanced economy consumers. Nonetheless, Eurozone consumption may still have been higher than it might have otherwise been had European consumers experienced the same oil price increases that U.S. consumers did. Financial markets may help accelerate incorporation of this dynamic into oil prices, instead of waiting for resulting demand increases to actually occur.[125]

Oil prices have risen significantly in terms of all major currencies, suggesting that changes in the value of the U.S. dollar are unlikely to have been an important driver of the oil price increases.[126]

Further, as shown in Figure 9, changes in the value of the U.S. dollar are dwarfed by changes in the price of oil. Also, it should be noted that the relationship may run two-ways—the price of oil may also affect the value of the U.S. dollar.[127] Because of the importance of oil to the U.S. economy and the size of U.S. oil imports, changes in the oil price impact economic activity and the U.S. trade deficit. These factors in turn affect the value of the U.S. dollar.[128] Finally, the U.S. dollar and the price of oil may move together at times because they are affected by some of the same drivers, namely economic growth rates, which differ by country.

A separate argument regarding the value of oil and the U.S. dollar is that oil exporters with a U.S. dollar-indexed exchange rate experience worsening terms of trade when the U.S. dollar depreciates—their currencies' values fall with the U.S. dollar. Many OPEC countries are indexed to the dollar. Some argue this in turn could make them more likely to target higher prices to offset the impact of a falling dollar, adjusting production policy accordingly.[129] Alternatively, it may be that the group targets the maximum sustainable price for oil, regardless of the value of the dollar.

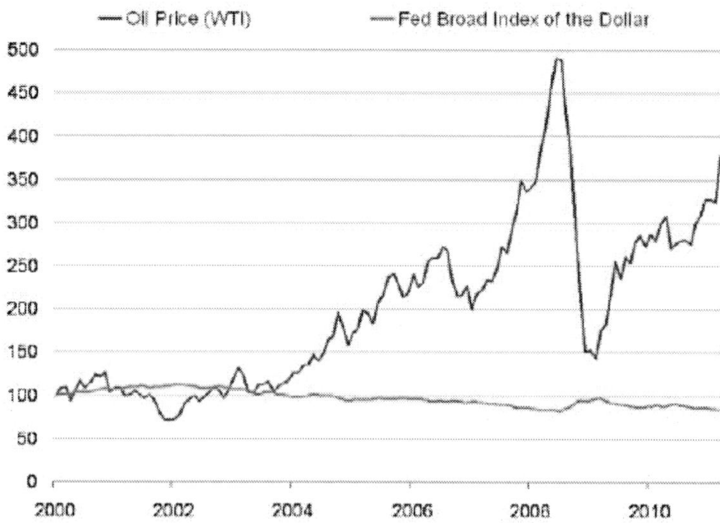

Source: EIA and Federal Reserve Board,
   http://www.federalreserve.gov/datadownload/.
Notes: The broad U.S. dollar index reflects the changing value of the dollar against the currencies of the United States' 26 largest trading partners, weighing the individual currencies in proportion to their trade with the United States. It is a more accurate measurement of the "value" of the U.S. dollar than any individual exchange rate. For more details, see
   http://www.federalreserve.gov/pubs/bulletin/2005/winter05_index.pdf.

Figure 9. Oil Price and the Dollar. Index, January 2000=100.

CONCLUSION

The oil market is globally integrated, but both supply and demand face constraints that can contribute to oil price fluctuations and oil prices trending higher in the future. Driven by future economic cycles, supply disruptions, and other global oil market developments, oil price fluctuations are expected to continue. Due to the price inelasticity of supply and demand, price swings have the potential to be large and disruptive to the economy. Over the long run, demand growth from emerging economies is expected to continue, as are controls on resource development by some oil-rich countries. The degree to which these trends drive up oil prices depends at least in part on how policy, investment, and new technology developments contribute to new sources of supply and greater demand efficiency. Additional supply sources are coming online, including in the United States, but they are technically more complex and may have some environmental draw backs, which represent higher costs than resources that have been available to date.

Policies that reduce U.S. oil demand or increase the supply of oil and oil alternatives can reduce economic risks posed when oil prices increase. By reducing the need for foreign oil, such policies reduce the drag on economic growth that can occur when oil price increases raise the unit cost of imports. Such policies can also contribute to lower oil prices by raising production or reducing consumption in the global oil market where prices are formed. Efficiency improvements, especially in transportation, may address risks most directly—limiting the need for oil insulates consumers and the economy regardless of what happens to prices. Raising the supply of oil or oil alternatives reduces the need for imports, but consumers still have to purchase fuel, the price of which can still be pushed up by market developments.

Supply and demand policies which require financial incentives today—for instance, subsidies for new fuel technologies or processes—represent short-term costs for long-term benefits. Regulatory efficiency or alternative fuels mandates may avoid up-front fiscal costs, but can still constitute a cost to consumers and businesses. Increasing conventional oil supplies through expanding access could add to federal revenues through royalties, but may raise risks of future environmental impacts such as oil spills and greenhouse gas emissions.

The costs, long lead times, or environmental risks associated with policies to address the market fundamentals of supply and demand can make policies that address non-fundamental issues an attractive option. For example, some favor further regulating speculation on oil futures contracts. However, it remains unclear how large or durable the impacts of such non-fundamental factors are. Nevertheless, increased transparency in financial markets could be one low-cost option that could improve price discovery and market reliability. Once fully implemented and enforced, Dodd-Frank may provide new information to policy makers and regulators that can clarify where there is additional opportunity for policy changes to improve the functioning of financial markets.

While various policy measures each have costs, so too does inaction. The relatively low, stable oil prices in the 1990s likely limited the incentive for policy makers and industry to prepare for the higher prices experienced since 2002. Given the scale and complexity of challenges posed by drivers of the oil price, it is unlikely that any single policy solution would address all price factors. A suite of policy options may be necessary to minimize the risks that oil prices pose to U.S. consumers and the economy. Given the long lead times for investment and technology development and deployment to provide benefits, commitments to such measures may need to persist through continued upward and downward oil price fluctuations.

End Notes

[1] This report will refer to WTI prices unless otherwise noted. Emerging differences between WTI and international crude oil benchmark prices is discussed in below in the box "WTI Price Depressed."

[2] One day closing prices. Data in this report come from the Energy Information Administration (EIA) unless otherwise noted, http://www.eia.gov.

[3] See http://www.eia.gov/totalenergy/data/annual/pecss_diagram.cfm. Excludes fuel ethanol in transportation.

[4] The remainder was largely refiner processing gains, volumetric gains that take place at the refinery when crude oil is processed into refined petroleum products such as gasoline.

[5] For more, see CRS Report R41765, *U.S. Oil Imports: Context and Considerations*, by Neelesh Nerurkar.

[6] U.S. Census Bureau, *U.S. International Trade in Goods and Services - Annual Revision for 2005*, June 9, 2006, Exhibit 8, http://www.census.gov/foreign-trade

[7] U.S. Census Bureau, *U.S. International Trade in Goods and Services - Annual Revision for 2010*, June 9, 2011, Exhibit 8, http://www.census.gov/foreign-trade

[8] Energy Information Administration, *Short Term Energy Outlook*, U.S. Department of Energy, May 10, 2011, http://www.eia.doe.gov/emeu/steo/pub/contents.html.

[9] Christopher Johnson, "Oil Tumbles As U.S. Downgrade Spurs Growth Concerns," *Reuters*, August 8, 2011. Concerns about growth in other countries also contributed to the fall in prices.

[10] See "Demand Elasticity and the Oil Market," p. 6.

[11] Richard Newell, *What's the Big Deal About Oil? How We Can Get Oil Policy Right*, Resources for the Future, 2006, p. 8, http://www.rff.org/rff/News/Features/upload/25007_1.pdf. Written prior to the start of 2007-2008 recession, which was also preceded by a run up in oil prices.

[12] Nigel Gault, "Oil Prices and the U.S. Economy: Some Rules of Thumb," *IHS Global Insight Inc.*, February 24, 2011. Estimates from other forecasters of the Gross Domestic Product (GDP) impact of a $10/bbl increase are similar, including those from the Federal Reserve according to reports (Robin Harding, "Oil Surge Puts Fragile US Recovery at Risk," *Financial Times*, February 24, 2011).

[13] The Federal Reserve may elect to raise interest rates if oil price increases lead to higher than desired inflation.

[14] This shift can be to oil production companies and—depending on oil industry cost inflation—to companies that provide the oil industry goods and services. This is discussed further below, see "Challenges to Oil Supply Growth."

[15] Nassir Khilji, *Economic Effects of High Oil Prices*, EIA, Annual Energy Outlook, Issues in Focus, 2006,
http://www.eia.gov/oiaf/aeo/otheranalysis/aeo_2006analysispapers/efhop.html.

[16] Matthew Higgins, Thomas Klitgaard, and Robert Lerman, *Recycling Petrodollars*, Federal Reserve Bank of New York, Current Issues in Economics and Finance, Volume 12, Number 9, December 2006, http://www.newyorkfed.org/ research/current_issues/ci12-9.pdf.

[17] Ibid. Authors point out that this financial flow from oil exporters to the United States has been indirect. Oil exporters have invested in other markets and the investments have been indirectly recycled to help finance the U.S. trade deficit.

[18] Lutz Kilian, Alessandro Rebucci, and Nikola Spatafora, "Oil Shocks and External Balances," *Research Seminar in International Economics*, no. 62 (March 29, 2007).

[19] Khilji, 2006.

[20] James Hamilton, "Causes and Consequences of the Oil Shock of 2007–08," *Brookings Papers on Economic Activity*, Spring 2009, pp. 39-40, http://www.nber.org/papers/w15002.

[21] Christof Rühl, "BP Statistical Review of World Energy," BP Plc., June, 2008.

[22] James D. Hamilton, "Understanding Crude Oil Prices," *The Energy Journal*, Vol. 30, No. 2, 2009, pp.189-195. Also see James D. Hamilton, "Fundamentals, speculation, and oil prices," http://www.econbrowser.com/archives/2011/08/ fundamentals_sp.html.·

[23] Joyce M. Dargay and Dermot Gately, "World oil demand's shift toward faster growing and less price-responsive products and regions," February, 2010, p. 29,
http://www.econ.nyu.edu/user/nyarkoy/ OilDemand_DargayGately_Feb2010.pdf.

[24] The fiscal and energy security implications of oil subsidies have led several governments, such as China, Russia, India, and Indonesia to start removing subsidies. For more see International Energy Agency, *Energy Subsidies: Getting Prices Right*, June 7, 2010, http://www.iea.org/files/energy_subsidies.pdf.

[25] Jonathan E. Hughes, Christopher R. Knittel, and Daniel Sperling, 2008. "Evidence of a Shift in the Short-Run Price Elasticity of Gasoline Demand," The Energy Journal, International Association for Energy Economics, vol. 29(1).

[26] For instance, oil prices were expected to level off at about $30 per barrel in the 2002 EIA International Energy Outlook (http://www.eia.gov/oiaf/archive/ieo02/pdf/world_oil.pdf) high price case, constrained by the availability of oil and oil alternatives at this price. This view was widely held, evidenced by similar forecasts from other sources also presented in that report.

[27] OPEC's role is discussed below.

[28] IHS/CERA, "IHS/CERA Upstream Capital Costs Index: Cost of Constructing New Oil and Gas Facilities Reaches New High," press release, May 14, 2008, http://www2.cera.com/news/details/1,2318,9487,00.html.

[29] Amy Meyers Jaffe and Ronald Soligo, *The International Oil Companies*, James A. Baker Institute for Public Policy, November 2007,
http://www.rice.edu/energy

[30] Bruce Bawks et al., *Performance Profiles of Major Energy Producers 2009*, EIA, February 2011, pp. 22-24, http://www.eia.gov/finance/performanceprofiles/pdf/020609.pdf.

[31] Lawrence Eagles, *Medium Term Oil Market Uncertainties*, Joint Transportation Research Center, Discussion Paper 2007-19, December 2007, http://www.internationaltransportforum.org/jtrc/discussionpapers/DiscussionPaper19.pdf.

[32] Barbara Saunders, "Upstream O&G Costs Rise Anew, IHS CERA Reports," *Rigzone*, July 1, 2011.

[33] Jaffe, p. 27, 2007.

[34] Countries do not all report their reserves in like terms, so these comparisons are difficult at best and should be interpreted as indicators of where known oil resources are concentrated. For more, see CRS Report R40872, *U.S. Fossil Fuel Resources: Terminology, Reporting, and Summary*, by Gene Whitney, Carl E. Behrens, and Carol Glover.

[35] CRS Report RL34137, *The Role of National Oil Companies in the International Oil Market*, by Robert Pirog.

[36] Quotas do not apply to production of natural gas liquids, which have been growing rapidly in many OPEC countries.

[37] Energy Information Administration, *Supply: OPEC*, What Drives Crude Oil Prices, http://www.eia.gov/finance/ markets/supply-opec.cfm.

[38] The OPEC Statute explains that the organization aims to safeguard the interests of its members, "devise ways and means of ensuring the stabilization of prices in international oil markets with a view to eliminating harmful and unnecessary fluctuations," provide steady income to producing countries, a steady supply of oil to consumers, and a fair return to investors in the petroleum industry. OPEC, *OPEC Statute*, http://www.opec.org/opec_web static_files_project/media/downloads/publications/OS.pdf.

[39] Robert McNally and Michael Levi, "A Crude Predicament: The Era of Volatile Crude Oil Prices," *Foreign Affairs*, July/July 2011.

[40] See "Refining and Crude Oil Prices" below.

[41] Only Russia and Saudi Arabia export more than 3 Mb/d. A number of oil exporters do export around 2 Mb/d, including Iraq, Iran, and Nigeria. Disruptions of any two of these, or disruptions in Saudi Arabia itself, could have dramatic impacts on the price of oil. This is particularly so if the disruptions were expected to result in prolonged supply outages as that would challenge the ability of the International Energy Agency (IEA) to cover the loss with strategic stockpiles. For more about strategic stockpile releases, see the section "June 2011 Strategic Petroleum Reserve Release" below.

[42] Iraq is a member of OPEC but does not currently participate in quota arrangements. For now Iraq, like non-OPEC countries, produces oil at the maximum possible capacity. In the foreseeable future, Iraq is a leading potential source of global supply growth. If and when it will rejoin OPEC quota arrangements remain unclear. For more on Iraq's plans to raise oil production capacity, see CRS Report RL34064, *Iraq: Oil and Gas Sector, Revenue Sharing, and U.S. Policy*, by Christopher M. Blanchard.

[43] According to EIA, Saudi Arabia supplied 10.5 Mb/d in 2010. Russia, the world's second largest producer in 2010, supplied 10.1 Mb/d. Saudi Arabia produced twice as much as OPEC's next largest producer, Iran (4.3 Mb/d).

[44] Cuts are apportioned to countries based on their production capacity. But some other OPEC countries, like Iran and Venezuela, despite being price hawks, are frequently suspected of not fully complying with cuts when they are announced. Saudi Arabia and other countries that comply with cuts may thus are responsible for a disproportionate amount of cuts.

[45] Robert L. Hirsch, Roger H. Bezdek, and Robert M. Wendling , *The Impending World Energy Mess* (Apogee Prime , 2010).

[46] For instance, see Peter Jackson, *The Future of Global Oil Supply: Understanding the Building Blocks*, IHS CERA, November 2009, http://press.ihs.com/press-release/energy-

[47] Modern peak oil theory proponents frequently cite the analysis of L. King Hubbert, who predicted that U.S. oil production would peak in 1970 on the basis of his geological analysis of U.S. oil fields. His followers have extrapolated Hubbert's work to claim that world oil production will soon peak, or may have already peaked. These claims were made throughout the 2000s when prices were increasing.

[48] EIA global oil production less biofuels.

[49] See the EIA's 2011 Annual Energy Outlook (p. 152, http://www.eia.gov/forecasts/aeo/pdf/0383(2011).pdf), OPEC's 2010 World Oil Outlook (p. 71, http://www.opec.org/opec_web/en/publications/340.htm), or IEA's 2010 World Oil Outlook (p. 113, http://www.iea.org/weo/).

[50] For instance, there are concerns that conventional ethanol from corn competes with food needs, and there are limits to how much ethanol can be blended into the fuel mix of the existing vehicle fleet. For more, see CRS Report R40445, *Intermediate-Level Blends of Ethanol in Gasoline, and the Ethanol "Blend Wall"*, by Brent D. Yacobucci. There are concerns about the financial and environmental costs of converting coal into liquid fuels. Gas-to-liquids also may face cost challenges depending on the price of the input natural gas and capital costs.

[51] For instance, the IEA expects OECD consumption to fall through 2030 (International Energy Agency, *World Energy Outlook*, November 9, 2010). The EIA AEO forecasts OECD consumption relatively flat through 2035 assuming current policies persist.

[52] BP, *2011 Statistical Review of World Energy*, June 2011, http://www.bp.com/statisticalreview.

[53] Pervin and Gertz Inc., *Study on Oil Refining and Oil Markets*, Prepared for the European Commission, 2008, p. 12, http://ec.europa.eu/energy

[54] A barrel is 42 gallons, but refining a barrel of crude oil yields slightly more than 42 gallons of products because these hydrocarbons expand in the refining process http://www.eia.gov/energyexplained/index.cfm?page=oil

[55] More recently, the spread has inverted—Maya crude has at times been more expensive than WTI. See the "WTI Price Depressed" below for an explanation of why WTI prices have been lower than international crude prices lately.

[56] U.S. Congress, House Committee on Energy and Commerce, *Gasoline: Supply, Price, and Specifications*, 109th Cong., 2nd sess., May 10, 2006, Serial No. 109-94 (Washington: GPO, 2006), p. 34.

[57] U.S. Government Accountability Office, Gasoline Markets: Special Gasoline Blends Reduce Emissions and Improve Air Quality, but Complicate Supply and Contribute to Higher Prices, GAO-05-421, June 15, 2005, http://www.gao.gov/ products/GAO-05-421.

[58] Damian Kennaby et al., *WTI Disconnect: Fantasy or Reality?*, Purvin and Gertz, March 15, 2011, http://www.purvingertz.com/content/articles/WTI%20Disconnect.pdf.

[59] Ibid.

[60] Adam Sieminski and Xiao Fu, "Global Commodities Daily" Deutsche Bank, June 15, 2011.

[61] Aaron Clark, "ConocoPhillips Not Interested in Reversing Seaway Pipeline," *Bloomberg*, February 15, 2011.

[62] U.S. Department of State, *Final Environmental Impact Statement* , Fact Sheet, August 26, 2011, http://www.state

[63] Christof Rühl, "BP Statistical Review of World Energy," BP Plc., June 2009.

[64] For more on this, see CRS Report R41683, *Middle East and North Africa Unrest: Implications for Oil and Natural Gas Markets*, by Michael Ratner and Neelesh Nerurkar.

[65] House Committee on Natural Resources (Minority), "Markey Pushes House Support for Oil Reserve Option," press release, March 7, 2011, http://democrats.

[66] 42 U.S.C. § 6241. The 'international energy program' is a reference to the International Energy Agency.

[67] Timothy Gardner and Ayesha Rascoe, "Analysis: One More Oil Spike May Push Obama to Tap SPR," *Reuters*, March 14, 2011.

[68] International Energy Agency, "IEA makes 60 million barrels of oil available to market to offset Libyan disruption," press release, June 23, 2011, http://www.iea.org/press/pressdetail.asp?PRESS_REL_ID=418.

[69] Part of the difficulty with rising crude production from the Middle East is that it tends to be heavier and more sour than relatively light, sweet Libyan exports. The U.S. SPR released light, sweet crude.

[70] U.S. Senate Committee on Energy and Natural Resources, "On SPR Oil Release, State of Chairman Jeff Bingaman," press release, June 23, 2011, http://energy PressRelease_id=9de08ce7-7838-487a-a324-d2bd7a2a193a.

[71] U.S. House Energy and Commerce Committee, "Upton Statement on President's Decision to Tap Strategic Petroleum Reserve," June 23, 2011, http://energycommerce.house.gov/News/PRArticle.aspx?NewsID=8745

[72] Emergency releases were also made in 1990-1991 after Iraq's invasion of Kuwait, and again in 2005 after Hurricanes Katrina and Rita. There have been non-emergencies sales of U.S. SPR crude and exchanges of SPR crude, which are detailed by the Department of Energy at http://www.fe.doe.gov/programs/reserves

[73] Muriel Boselli, "Germany, Italy may resist second IEA oil release," *Reuters*, July 15.

[74] U.S. Department of Transportation, National Highway Traffic Safety Administration, Summary of Fuel Economy Performance, April 20, 2011, http://www.nhtsa.gov/fuel-economy.

[75] Portions of this paragraph taken from CRS Report R40645, *U.S. Offshore Oil and Gas Resources: Prospects and Processes*, by Marc Humphries, Robert Pirog, and Gene Whitney.

[76] Lease sales 216, 218, 220, and 222. The Administration's Revised 2007-2012 Leasing Program does currently have scheduled lease sales 216, 218, and 222, which are in the Central or Western Gulf of Mexico.

[77] EIA analysis in the AEO 2011 suggests that new lease sales in the OCS decrease the price of light, low sulfur crude oil by between $1.47 and $5.97/bbl in 2035, depending on what assumptions are made about OCS resources. (http://www.eia.gov/oiaf/aeo/tablebrowser/) In 2008, EIA has studied the impact of developing oil in ANWR, forecasting that it may reduce oil prices in 2027 by between $0.41 and $1.44/bbl, again depending on resource assumptions. Both analyses did find that opening up new areas would have a significant impact on domestic production and trade balances. Regarding the price forecasts, EIA concludes: "Changes in domestic oil production tend to have only a modest impact on crude oil and petroleum product prices, because any change in domestic oil production is diluted in the world oil market." (http://www.eia.gov/forecasts/aeo/pdf/0383(2011).pdf)

[78] Department of Transportation and the Environmental Protection Agency, "2017–2025 Model Year Light-Duty Vehicle GHG Emissions and CAFE Standards: Supplemental Notice of Intent," 76 *Federal Register* 48759, August 9, 2011.

[79] National Highway Traffic Safety Administration and the Environmental Protection Agency, *Paving the Way Toward Cleaner, More Efficient Trucks*, August 9, 2011, http://www.nhtsa.gov/staticfiles/rulemaking/pdf/cafe/ Factsheet.08092011.pdf.

[80] For example, H.R. 1861.

[81] For example, S. 670.

[82] For example, S. 612.

[83] For example, H.R. 910 or S. 228.

[84] David Shepardson and Christina Rogers, "GM's Akerson pushing for higher gas taxes," *The Detroit News*, June 7, 2011. For more on transportation funding, see CRS Report R41490, *Surface Transportation Funding and Finance*, by Robert S. Kirk and William J. Mallett.

[85] Ron Gecan and Rob Johansson, *Using Biofuel Tax Credits to Achieve Energy and Environmental Goals*, Congressional Budget Office, July 2010, http://www.cbo.gov/ftpdocs/114xx/doc11477/07-14-Biofuels.pdf.

[86] CRS Report R40445, *Intermediate-Level Blends of Ethanol in Gasoline, and the Ethanol "Blend Wall"*, by Brent D. Yacobucci.

[87] For example, H.R. 1075 or provisions in S. 782.

[88] For example, H.R. 500 and S. 734 (which also includes provisions for natural gas fueled, hybrid, and advanced internal combustion engine vehicles).

[89] For example, H.R. 1380.

[90] For example, H.R. 909.

[91] U.S. Department of Energy, *President Requests $520.7 Million for Fossil Energy Programs*, Fossil Energy Techline, February 14, 2011, http://fossil.energy

[92] Though even within the OECD the timeliness of data varies, with data from the United States and Japan being more up to date than European statistics.

[93] Chen Aizhu, "Q+A- Why Is China Not Releasing Oil Inventory Data Yet?" *Reuters*, November 8, 2010.

[94] International Energy Forum, *Concluding Statement by Host Country Mexico and Co-hosts Germany and Kuwait*, March 31, 2010, http://www.ief.org/Events/Pages/12thIEFMinisterial.aspx. Note, the IEF here is likely referring to investors broadly, including those who make energy production and consumption related capital investments.

[95] G-20, *The G-20 Toronto Summit Declaration*, July 27, 2010, p. 3, http://www.g20.org/Documents/ g20_declaration_en.pdf. Also, see footnote 24.

[96] U.S. China Oil and Gas Industry Forum, *Background*, 2010, http://www.uschinaogf.org/background.html.

[97] U.S.-China Clean Energy Center, *Joint Work Plan for Collaborative Research on Clean Vehicles*, http://www.uschina-cerc.org/Clean_Vehicles.html.

[98] Attorney General Eric Holder, *Memorandum to the Financial Fraud Enforcement Task Force*, Department of Justice, April 21, 2011, http://www.justice.gov/ag/AG_Memo_to_FFETF-Gas_Prices.pdf.

[99] Examples of hedgers include oil companies, refiners, petrochemical companies, utilities, and airlines.

[100] For example, if the futures price for delivery of oil three months from now was $120, and the current spot market price was $100, arbitrageurs could sell the future and buy physical oil to settle the contracts when they expired. The differential of $20 per barrel (minus the cost of storage) would be riskless profit. Arbitrage transactions would lower the futures price and raise the physical price, forcing the prices back together.

[101] For more background, see CRS Report R40646, *Derivatives Regulation and Recent Legislation*, by Mark Jickling and Rena S. Miller, and The Mechanics of the Derivatives Markets by the International Energy Agency (http://omrpublic.iea.org/special_sup_apr11.pdf).

[102] The chart actually understates the proportion of speculators, because some traders classified as "commercial" are hedging speculative contracts with others, rather than physical market price exposure. More recent CFTC data disaggregate swap dealers from producer/merchant hedgers, but these data are not available before 2006.

[103] The classic theoretical formulation of this insight is Milton Friedman's: speculators must on average *decrease* price volatility, because they go out of business unless they can they buy low and sell when prices are high, thus smoothing out the peaks and valleys of price trends. See *Essays in Positive Economics* (Chicago, University of Chicago Press, 1953), p. 175.

[104] CRS Report R41902, *Hedge Fund Speculation and Oil Prices*, by Mark Jickling and D. Andrew Austin.

[105] U.S. Congress, Senate Committee on Energy and Natural Resources, Subcommittee on Energy, *Speculative Investment in Energy Markets*, Committee Print, 110th Cong., 2nd sess., September 16, 2008, pp. 10-12.

[106] Interagency Task Force on Commodity Markets, "Interim Report on Crude Oil Markets," July 2008, http://www.cftc.gov/ucm/groups/public/@newsroom/documents/ file/itfinterim reportoncrudeoil0708.pdf.

[107] Marco J. Lombardi and Ine Van Robays, *Do Financial Investors Destabilize the Oil Price*, European Central Bank, Work Paper Series Number 1346, June 2011, http://www.ecb.int/pub/pdf/scpwps/ecbwp1346.pdf.

[108] CFTC, Complaint for Injunctive and Other Equitable Relief and Civil Monetary Penalties Under the Commodity Exchange Act: U.S. Commodity Futures Trading Commission v. Parnon Energy Inc., Arcadia Petroleum LTD, Arcadia Energy (Suisse) SA, Nicholas J. Wildgoose and James T. Dyer, May 24, 2011, http://www.cftc.gov/ucm/groups/public/ @lrenforcementactions/documents/legalpleading/enfparnoncomplaint052411.pdf.

[109] As noted above, the vast majority of oil futures are settled in cash without physical delivery. But traders with contracts to buy oil have the right to take physical delivery. If they choose to do so, those traders left with open contracts to sell oil must actually deliver oil of the quantity, quality, and at the location specified in the contracts. If the number of traders "standing for delivery" rises unexpectedly, the conditions for a "squeeze" are present.

[110] Commodity Exchange Act, section 4a(a), codified at 7 USC §6a.

[111] The CFTC proposed rules regarding position limits in January 2011. "Position Limits for Derivatives," *Federal Register*, v. 76, Jan. 26, 2011, p. 4752.

[112] For oil, a reportable position is 350 futures contracts.

[113] In 2008, the CFTC found that "[t]o date, there is no statistically significant evidence that the position changes of any category or subcategory of traders systematically affect prices. This is to be expected in well-functioning markets." Interagency Task Force on Commodity Markets, "Interim Report on Crude Oil Markets, p. 31. However, there are differing views within the CFTC on the impact of speculation—see, e.g., Commissioner Bart Chilton, "Caging the Financial Cheetahs," Speech to American Soybean Association Legislative Forum, Washington, DC, July 12, 2011.

[114] For more see CRS Report R41398, *The Dodd-Frank Wall Street Reform and Consumer Protection Act: Title VII, Derivatives*, by Mark Jickling and Kathleen Ann Ruane.

[115] CRS Report RS22236, *Gasoline Price Increases: Federal and State Authority to Limit "Price Gouging"*, by Adam Vann and Kathleen Ann Ruane.

[116] Alabama, Arkansas, California, Connecticut, Florida, Georgia, Hawaii, Idaho, Illinois, Indiana, Iowa, Kansas, Kentucky, Louisiana, Maine, Massachusetts, Michigan, Mississippi, Missouri, New Jersey, New York, North Carolina, Oklahoma, South Carolina, Tennessee, Texas, Utah, Virginia, West Virginia, and Wisconsin—the District of Columbia, Guam, and the Northern Mariana Islands. Other states may also exercise authority under general deceptive trade practice laws depending on the nature of the state law and the specific circumstances in which price increases occur.

[117] For more information on the applicability of the federal antitrust laws, see CRS Report RS22262, *"Price Gouging," the Antitrust Laws, and Vertical Integration in the Petroleum Industry: How They Are Related*, by Janice E. Rubin.

[118] Federal Trade Commission, *Report on Spring/Summer 2006 Nationwide Gasoline Price Increase*, August 2007, p. 26, http://www.ftc.gov/reports/gasprices06/P040101Gas06increase.pdf .

[119] Since 1981, the FTC has alleged that 15 proposed petroleum mergers would have resulted in significant reductions in competition and harmed consumers in one or more relevant markets. Four of the mergers were abandoned or blocked as a result of FTC or court action. In the other 11 cases, the FTC required the merging companies to divest substantial assets in the markets where competitive harm was likely to occur (http://www.ftc.gov/ftc/oilgas/enf_merge.htm).

[120] Federal Trade Commission, *FTC Releases Report on its "Investigation of Gasoline Price Manipulation and Post-Katrina Gasoline Price Increases,"* May 22, 2006, http://www.ftc.gov/opa/2006/05/katrinagasprices.shtm. Also Federal Trade Commission, *Report on Spring/Summer 2006 Nationwide Gasoline Price Increases*, 2007, p. 25, http://www.ftc.gov/reports/gasprices06/P040101Gas06increase.pdf.

[121] Section 811 of Subtitle B of Title VIII, P.L. 110-140.

[122] 16 C.F.R. Part 317, http://www.ftc.gov/os/2009/08/P082900mmr_finalrule.pdf.

[123] Similar provisions are also included in H.R. 1748, the Taxpayer and Gas Price Relief Act of 2011

[124] Frank Verrastro et al., *"NOPEC" Legislation and U.S. Energy Security*, Center for Strategic and International Studies, Critical Questions , June 14, 2011, http://csis.org/publication/nopec-legislation

[125] CRS Report RL34686, *The U.S. Trade Deficit, the Dollar, and the Price of Oil*, by James K. Jackson.

[126] U.S. Congress, Senate Committee on Banking, Housing, and Urban Affairs, *Semiannual Monetary Policy Report to the Congress*, 112th Cong., 1st sess., March 1, 2011.

[127] U.S. Congress, Senate Committee on Banking, Housing, and Urban Affairs, *Semiannual Monetary Policy Report to the Congress*, 110th Cong., 2nd sess., July 15, 2008.

[128] Jackson, p. 20-21.

[129] Jackson, p. 5-6.

In: Oil Prices   ISBN: 978-1-61942-485-2
Editors: J. C. Mullen and B. M. Lynn   © 2012 Nova Science Publishers, Inc.

Chapter 3

UNITED STATES TRADE DEFICIT AND THE IMPACT OF CHANGING OIL PRICES[*]

James K. Jackson

SUMMARY

Petroleum prices have risen sharply since September 2010, at times reaching more than $112 per barrel of crude oil. Although this is still below the $140 per barrel price reached in 2008, the rising cost of energy was dampening the rate of growth in the economy during the first half of 2011. While the price of oil has increased sharply, the volume of oil imports, or the amount of oil imported, has decreased slightly. Overall resistance by market demand to changes in oil prices reflect the unique nature of the demand for oil and an increase in economic activity that has occurred since the worst part of the economic recession in 2009. Turmoil in the Middle East was an important factor causing petroleum prices to rise sharply in the first four months of 2011, which could add as much as $100 billion to the total U.S. trade deficit in 2011. The increase in energy import prices is pushing up the price of energy to consumers and could spur some elements of the public to pressure the 112th Congress to provide relief to households that are struggling to meet their current

---

[*] This is an edited, reformatted and augmented version of a Congressional Research Service publication, CRS Report for Congress RS22204, from www.crs.gov, dated September 14, 2011.

expenses. With oil prices rising to over $100 per barrel in early 2011, the International Energy Agency cautioned that the rising price of oil was becoming a threat to the global economic recovery. This report provides an estimate of the initial impact of the changing oil prices on the nation's merchandise trade deficit.

BACKGROUND

According to data published by the Census Bureau of the Department of Commerce,[1] the prices of petroleum products during the first five months of 2011 rose sharply, generally rising faster than the change in demand for those products, similar to the experience in 2008. In 2008, however, after reaching nearly $140 per barrel, prices fell at a historic rate.[2] After falling each month between August 2008 and February 2009, average petroleum prices reversed course and rose by 85% between February and December 2009, climbing to nearly $80 per barrel at times.

In 2010, petroleum prices reached a peak average price of about $77 per barrel in April before falling to around $72 per barrel in July 2010. Average prices dropped from May to July, one of only three times average monthly petroleum prices have declined since January 2009. In December 2010, petroleum import prices averaged nearly $80 per barrel and continued to increase, reaching over $112 per barrel at times in March, April, and May 2011.

Oil futures contracts indicate, however, that crude oil prices are expected to peak at the $100 per barrel range before falling to under $90 per barrel by mid-fall 2011. Turmoil in the Middle East, natural disasters, and hurricanes, however, could have a significant impact on the course of oil prices for the foreseeable future. As a result of changing petroleum prices, the price changes in imported energy-related petroleum products worsened the U.S. trade deficit in 2006-2008, 2010, and likely will again in 2011.[3] *Energy-related petroleum products* is a term used by the U.S. Census Bureau that includes crude oil, petroleum preparations, and liquefied propane and butane gas. Crude oil comprises the largest share by far within this broad category of energy-related imports.

In 2009, the slowdown in the rate of growth in the U.S. economy reduced the amount of energy the country imported and helped push down world energy prices. Since then, economic growth has improved, energy imports have increased, and energy prices have risen.

Table 1. Summary Data of U.S. Imports of Energy-Related Petroleum Products, Including Oil (not seasonally adjusted)

	January - July					
	2010		2011			
	Quantity (millions of barrels)	Value ($ billions)	Quantity (millions of barrels)	% change 2010 to 2011	Value ($ billions)	% change 2010 to 2011
Total energy-related petroleum products	2,506.2	$188.4	2,451.4	-2.2%	$245.4	30.3%
Crude oil	1,983.8	$147.3	1,933.4	-2.5%	$190.0	29.0%
	January through December					
	2010		2011			
	(Actual values)		(Estimated values)			
	Quantity (millions of barrels)	Value ($ billions)	Quantity (millions of barrels)	% change 2010 to 2011	Value ($ billions)	% change 2010 to 2011
Total energy-related petroleum products	4,279.5	$323.8	4,186.0	-2.2%	$421.8	30.3%
Crude oil	3,377.7	$252.2	3,291.3	-2.5%	$325.3	29.0%

Source: U.S. Department of Commerce, U.S. Census Bureau, Report FT900, *U.S. International Trade in Goods and Services*, Table 17, September 8, 2011.

Note: Estimates for January through December 2011 were developed by CRS from data in January-July, 2011, and data through 2010 published by the Census Bureau using a straight line extrapolation.

In isolation from other events, lower energy prices tend to aid the U.S. economy, which makes it a more attractive destination for foreign investment. Such capital inflows, however, place upward pressure on the dollar against a broad range of other currencies.

To the extent that the additions to the merchandise trade deficit are returned to the U.S. economy as payment for additional U.S. exports or to acquire such assets as securities or U.S. businesses, the U.S. trade deficit could be mitigated further.

Summary data from the Census Bureau for the change in the volume, or quantity, of energy-related petroleum imports and the change in the price, or the value, of those imports for 2010 and estimated values for 2011 are presented in Table 1. The data indicate that during 2010, the United States imported about 4.3 billion barrels of energy-related petroleum products, valued at $323 billion.

On average, energy-related imports for 2010 were up 0.3% in volume terms from the average in 2009 and cost an average of 31% more than similar imports during the same period in 2009. These data demonstrate that U.S. demand for oil imports is highly resistant to changes in oil prices. According to various studies, U.S. demand for oil is correlated more closely to U.S. per capital income than to changes in oil prices.[4]

Estimates for 2011 indicate that with the average price of around $100 per barrel, U.S. imported petroleum costs could rise by $90 billion in 2011 to reach $414 billion.

The data also indicate that in 2009, the quantity of energy-related petroleum imports fell by 4.0% compared with the comparable period in 2008; crude oil imports also fell by 2.7% from the same period in 2008.

Year-over-year, the average value of energy-related petroleum products imports fell by 44% in 2009, while the average value of crude oil imports fell by 45%. As Figure 1 shows, imports of energy-related petroleum products can vary sharply on a monthly basis. In 2010, imports of energy-related petroleum products averaged about 356 million barrels per month.

In value terms, energy-related imports fell from a total value of $439 billion in 2008 to $245 billion in 2009, or a decrease of 44%, to account for about 16% of the value of total U.S. merchandise imports. Energy prices rose sharply in 2007 and continued rising from January through July 2008, not following previous trends of falling during the winter months.

The cost of U.S. imports of energy-related petroleum products rose from about $17 billion per month in early 2007 to $53 billion a month in July 2008, but fell to $13.6 billion a month in February 2009, reflecting a drop in the price and in the volume of imported oil. The average price of imported oil in July 2011 was $104.3, down from an average of $108.7 in May 2011, but up 47% over the average price per barrel of $72 in July 2010.

As Figure 2 shows, the value of total energy imports (reflecting the change in the amount of imports and the change in the price of those imports) in July 2011 fell 5.6% from June 2011 to reach $37.2 billion, up nearly 32% from the total value of energy imports in July 2010, as indicated in Table 2.

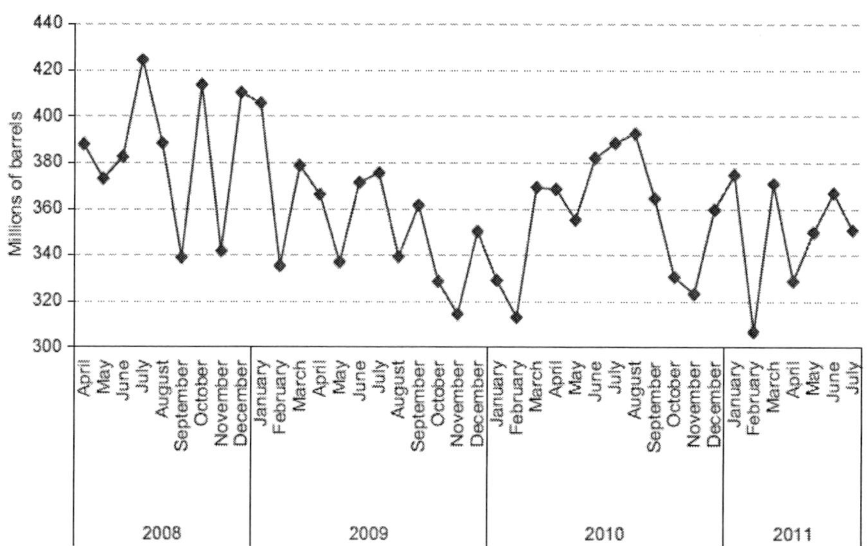

Source: Department of Commerce.

Figure 1. Quantity of U.S. Imports of Energy-Related Petroleum Products.

As a result of the drop in the overall value of energy-related imports in 2009, the trade deficit in energy-related imports amounted to $204 billion, down by nearly half from the $386 billion recorded in 2008, and accounted for 40% of the total U.S. trade deficit of $517 billion for the year.

In 2010, the rise in oil prices, year over year, combined with a slight increase in energy imports, pushed up the overall value of energy imports, which accounted for 41% of the total merchandise trade deficit. In July 2011, the share of the U.S. trade deficit arising from energy imports was 42%, down from the 44% recorded in June 2011.

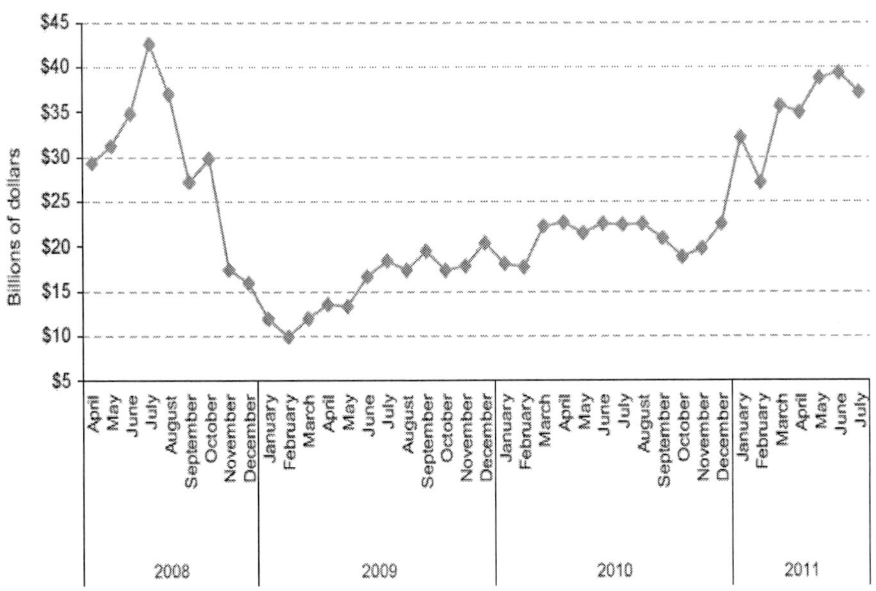

Source: Department of Commerce.

Figure 2. Value of U.S. Imports of Energy-Related Petroleum Products.

Crude oil comprises the largest share of energy-related petroleum products imports. According to Census Bureau data,[5] imports of crude oil fell from an average of 9.8 million barrels of crude oil imports per day in 2008 to an average of 9.1 million barrels per day in 2009, or a decrease of 7%.

In July 2011, such imports averaged 9.1 million barrels per day, or a decrease of about 7% from the volume of such imports recorded in June 2011. From January 2008 to June 2008, the average price of crude oil increased from $84 per barrel to $117 per barrel, or an increase of 39%. As a result, the value of U.S. crude oil imports rose from about $27 billion a month in January 2008 to $35 billion a month in June 2008.

In 2010, crude oil imports averaged 281 million barrels per month at an average value of $21 billion a month. Oil import prices in 2010 rose from about $74 per barrel in January 2010 to an average of $79.78 in December 2010.

As shown in Figure 3, oil import prices rose steadily between September 2010 and May 2011, before falling in June and July 2011.

Table 2. U.S. Imports of Energy-Related Petroleum Products, Including Crude Oil (not seasonally adjusted)

Period	Total energy-related petroleum products[a] Quantity (millions of barrels)	Value ($ billions)	Crude oil Quantity (millions of barrels)	Thousands of barrels per day (average)	Value ($ billions)	Unit price (dollars)
			2010			
Jan.-Dec.	4,279.5	$323.8	3,377.7	9,252	$252.2	$74.67
Jan.-July	2,506.2	188.4	1,983.8	9,357	147.3	74.25
January	328.8	24.7	244.8	7,897	18.1	73.95
February	312.9	23.0	242.7	8,669	17.7	72.96
March	368.5	27.7	298.5	9,628	22.2	74.32
April	369.6	28.9	294.9	9,829	22.7	77.13
May	357.5	27.8	281.6	9,082	21.7	76.95
June	383.9	28.1	313.3	10,443	22.7	72.39
July	385.2	28.2	308.0	9,937	22.2	72.09
August	392.9	29.2	307.1	9,905	22.6	73.47
September	367.0	26.8	292.6	9,754	21.2	72.33
October	329.0	24.9	252.2	8,135	18.7	74.21
November	323.7	25.2	257.9	8,596	19.8	76.81
December	360.7	29.3	283.6	9,147	22.6	79.79
			2011			
Jan.-July	2,451.4	245.4	1,933.4	9,120	190.0	98.28
January	375.3	32.2	290.7	9,376	24.5	84.34
February	307.3	27.2	242.4	8,656	21.1	87.17
March	371.4	35.7	295.1	9,520	27.7	93.76
April	329.2	35.0	252.2	8,408	26.0	103.18
May	350.7	38.8	275.3	8,879	29.9	108.70
June	366.8	39.4	296.7	9,889	31.4	106.00
July	350.7	37.2	281.1	9,067	29.3	104.27

Source: U.S. Department of Commerce, U.S. Census Bureau, Report FT900, U.S. International Trade in Goods and Services, Table 17, September 8, 2011.

a. Energy-related petroleum products is a term used by the Census Bureau and includes crude oil, petroleum preparations, and liquefied propane and butane gas.

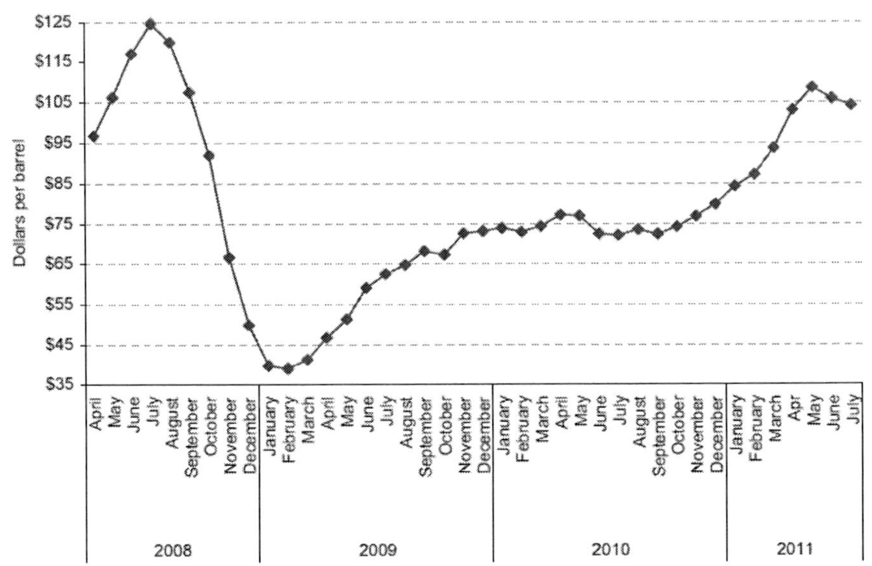

Source: Department of Commerce.

Figure 3. U.S. Import Price of Crude Oil.

Data for 2009 indicate that a number of factors, primarily the economic recession, had a large impact on pushing down oil prices in the first three months. As economic growth picked up, the higher demand for oil tended to raise pressure on oil prices, which rose through the end of the year. The rise in oil prices and an increase in the volumes of oil imports during the period combined to raise the overall cost of imported energy. At times, crude oil traded for nearly $148 per barrel in July 2008, indicating that the cost of energy imports would have a significant impact on the overall costs of U.S. imports and on the size of the U.S. trade deficit. Since those record prices, the price per barrel of imported crude oil fell to under $40 per barrel at times in January and February 2009. For the year 2009, the imported volume of energy-related petroleum products fell by 44% compared with 2008, due in large part to a slowdown in economic activity. At an average price of $56 per barrel in 2009, compared with an average price of $95 per barrel in 2008, energy-related imports fell by nearly $130 billion as a component in the overall U.S. trade deficit. For 2010, the total cost of energy imports rose to $323 billion at an average price of $75 per barrel and accounted for 41% of the annual trade deficit. Estimates for 2011 indicate that at an average price of imported energy of about $100 per barrel, the total cost of energy imports could rise to $400 to $425 billion, or nearly $100 billion more than the cost of energy imports in 2010.

ISSUES FOR CONGRESS

The rise in the prices of energy imports experienced since early 2010 through May 2011 could have a significant impact on the annual U.S. trade deficit in 2011, should those price increases stick, or run even higher. The rise in energy prices experienced during the first half of 2011 could affect the U.S. rate of inflation and could have a slightly negative impact on the rate of economic growth in 2011. Various factors, dominated by the political turmoil in the Middle East and the rate of economic growth in Asia and other developing economies, have combined to push up the cost of energy imports, which will have a slightly negative impact on the pace of the economic recovery. The pace of economic growth, however, seems to have stalled during the second quarter of 2011, which could have an important effect on both the levels of oil imports and on the price of such imports if economic growth remains listless. Typically, energy import prices have followed a cyclical pattern as energy prices rose in the summer and declined in the winter. The slowdown in the rate of economic growth in the United States and elsewhere in 2009 sharply reduced the demand for energy imports and caused oil prices to tumble from the heights they reached in July 2008. An important factor that often affects crude oil prices is the impact Atlantic hurricanes have on the production of crude oil in the Gulf of Mexico.

The return to a positive rate of economic growth in 2010 placed upward pressure on the prices of energy imports and contributed to the nation's merchandise trade deficit. Some of the impact of this deficit could be offset if some of the dollars that accrue abroad are returned to the U.S. economy through increased purchases of U.S. goods and services or through purchases of such other assets as corporate securities or acquisitions of U.S. businesses. Some of the return in dollars likely will come through sovereign wealth funds, or funds controlled and managed by foreign governments, as foreign exchange reserves boost the dollar holdings of such funds. Such investments likely will add to concerns about the national security implications of foreign acquisitions of U.S. firms, especially by foreign governments, and to concerns about the growing share of outstanding U.S. Treasury securities that are owned by foreigners.

Social turmoil in the Middle East created uncertainty in the oil markets through the first half of 2011 and was a major factor in rising oil prices. The duration and intensity of the turmoil likely will continue to be the most important factor driving oil prices. As was the case in 2008, high and sustained oil prices likely will have a detrimental effect on the pace of economic growth in many parts of the world. It is possible for the economy to adjust over the long term to the higher prices of energy imports by improving its energy efficiency, finding alternative sources of energy, or searching out additional supplies of energy. Higher oil prices may well cause consumers to increase pressure on Congress to assist in this process. For Congress, the increase in the nation's merchandise trade deficit could add to existing inflationary pressures and complicate efforts to reduce the governments' budget deficit and to stimulate the economy should the rate of economic growth stall. In particular, Congress, through its direct role in making economic policy and its oversight role over the Federal Reserve, could face the dilemma of rising inflation, which generally is treated by raising interest rates to tighten credit, and a slow rate of economic growth, which is usually addressed by lowering interest rates to stimulate investment. A sharp rise in the trade deficit may also add to pressures for Congress to examine the causes of the deficit and to address the underlying factors that are generating that deficit. In addition, the rise in prices of energy imports could add to concerns about the nation's reliance on foreign supplies for energy imports and add impetus to examining the nation's energy strategy.

End Notes

[1] U.S. Department of Commerce, U.S. Census Bureau, Report FT900, *U.S. International Trade in Goods and Services,* Table 17, September 8, 2011. The report and supporting tables are available at http://www.census.gov/foreign-trade Press-Release/currentjress_release/ ftdpress.pdf.

[2] For information about the causes of the run up in oil prices see Hamilton, James, Causes and Consequences of the Oil Shock of 2007-2008, *Brookings Papers on Economic Activity,* Spring 2009.

[3] For additional information about U.S. oil imports see CRS Report R41765, *U.S. Oil Imports: Context and Considerations*, by Neelesh Nerurkar.

[4] Hamilton, Causes and Consequences of the Oil Shock of 2007-2008; *World Economic Outlook,* Chapter 3, International Monetary Fund, April 2011. According to the IMF, for developed economies, a 10% increase in oil prices is estimated to result in a 0.2% decrease in oil consumption, but a 10% increase in income leads to a 6.8% increase in oil consumption.

[5] Report FT900, *U.S. International Transactions in Goods and Services*, Table 17, September 8, 2011.

In: Oil Prices   ISBN: 978-1-61942-485-2
Editors: J. C. Mullen and B. M. Lynn   © 2012 Nova Science Publishers, Inc.

Chapter 4

THE UNITED STATES TRADE DEFICIT, THE DOLLAR, AND THE PRICE OF OIL[*]

James K. Jackson

SUMMARY

Rapid changes in the price of oil and the impact of such price changes on economies around the globe have attracted considerable attention. In mid-2008 as the price of oil rose to unprecedented heights and then dropped sharply, the international exchange value of the dollar fell and then rose relative to a broad basket of currencies. For some, these two events seem to indicate a cause and effect relationship between changes in the price of oil and changes in the value of the dollar. Despite common perceptions that there is a direct cause and effect relationship between changes in the international exchange value of the dollar and the price of oil, an analysis of data during recent periods indicates that changes in the price of oil are driven by changes in the demand for oil that is different from the supply of oil, rather than changes in the value of the dollar. The rapid increase in oil prices in early 2011 reflects rising demand for oil and other commodities and uncertainty in global markets keyed to political turmoil in North Africa and the Middle East.

This report analyzes the relationship between the dollar and the price of oil and how the two might interact. While the data do not support a strong cause and effect relationship between the value of the dollar and

---

[*] This is an edited, reformatted and augmented version of a Congressional Research Service publication, CRS Report for Congress RL34686, from www.crs.gov, dated May 26, 2011.

the price of oil, there likely are various channels through which changes in the price of oil and in the value of the dollar may be indirectly correlated. The data also indicate that an increase in the demand for crude oil that exceeded the increase in the supply of oil and a laggardly pace in oil production capacity likely are among the main factors behind the sharp run up in the price of oil that occurred during the first seven months of 2008. Changes in oil prices also affect the U.S. trade deficit. That impact of oil prices on the trade deficit lessened in 2009 and 2010 as the price of oil plummeted and as a drop in economic activity reduced demand for oil imports. A more rapid pace of economic growth in many of the developing economies in 2011 exerted upward pressure on oil prices that will translate into a higher U.S. trade deficit in 2011. This report provides an assessment of the impact a range of prices of imported oil could have on the U.S. trade deficit.

OVERVIEW

To most observers, it seems apparent that the rise in the price of oil[1] and the decline in the exchange value of the dollar often are interconnected events, or that there is some cause and effect relationship between the two.[2] Since oil is priced in dollars, this line of reasoning goes, as the exchange value of the dollar declines, the purchasing power of oil producers also falls, which, in turn, prods oil producers to reduce their supplies to the market in order to push up the market price of oil and restore their purchasing power. This line of thinking is not unreasonable, considering various incidents, most notably 1973 and 1979, in which the price of oil rose sharply in response to actions taken by members of the Organization of Petroleum Exporting Countries (OPEC)[3] group of oil producers to increase the market price of oil. Indeed, OPEC's stated objective is to coordinate and unify petroleum policies among OPEC Countries, in order to secure "fair and stable prices for petroleum producers; an efficient, economic and regular supply of petroleum to consuming nations; and a fair return on capital to those investing in the industry."

After reaching nearly $147 per barrel in August 2008, the price per barrel of oil dropped to less than $40 per barrel by year-end 2008, before rising again through May 2011 to reach as high as $120 per barrel at times. In response to the drop in oil prices in 2008, OPEC announced cuts in production on three occasions: a cut of 500,000 barrels per day announced on September 1, 2008,[4] a cut of 1.5 million barrels per day announced on October 25, 2008, and a cut of 2.2 million barrels per day announced on December 17, 2008. In February 2011, Saudi Arabia increased oil production to calm market fears over political unrest in North Africa and the Middle East, particularly concerns over the impact political turmoil on Libya could have on oil production there.

An analysis of the data indicates that fluctuations in the price of oil that have been experienced since 2006 have not been driven primarily by a reduction in world supplies. Instead, the changes in oil prices reflect a number of factors, including the slow-paced growth in oil production; an increase in demand, most notably among the developing countries, that has outpaced the increase in supply; and more recently market concerns related to political turmoil. Changes in the international exchange value of the dollar, however, likely reflect a number of factors, including changes in the demand for and supply of capital within the U.S. economy, the relative rate of return on interest-sensitive assets, and expectations about the performance of the U.S. economy. At the same time, some observers have argued that oil market speculators played an important role in pushing up oil prices so quickly in 2008.[5] A report issued on September 11, 2008, by the Commodity Futures Trading Commission (CFTC), however, concluded that market speculators probably were not responsible for the rise in oil prices.[6]

While data on exchange rates and on oil prices do not support the case for a strong cause and effect relationship between the value of the dollar and the price of oil, there are a number of channels through which changes in the price of oil and changes in the value of the dollar may be indirectly correlated. In fact, an increase in the price of oil to offset the loss of purchasing power that is associated with a depreciation in the value of the dollar can spark a chain of events that could blunt or even nullify the rise in oil prices.

The pervasive nature of such commodities as oil, which serve as essential components in economic growth, means that changes in the prices of those commodities affect the prices of a broad range of goods, services, and economic activities.[7] Indeed, according to the Census Bureau, increases in the price of imported oil were a major factor in rising consumer prices in the United States in the first six months of 2008. Similarly, rising oil prices in late 2010 and early 2011 have pushed up the prices of other commodities. Rising consumer and commodity prices undermine the exchange value of the dollar relative to other currencies and reduce the real incomes of consumers, which can lead to a lower rate of economic growth. Slower economic growth, in turn, lowers the demand for oil, thereby putting downward pressure on the price of oil, as occurred in 2009.[8] Expectations about future economic growth and, therefore, about the demand for crude oil, also can affect a broad range of investment decisions that might affect expectations about the value of the dollar. The interaction between the price of oil and the value of the dollar is complicated further by the way changes in the price of oil can affect the economic performance of other nations and, therefore, have an impact on their respective currencies.[9]

Oil is also highly unique, because it has few substitutes. The long lag times that are required to bring new fields on line and the prominent role oil-related products occupy in the economy mean that oil defies come common notions of market supply and demand. For consumers, there are few short-term available substitutes when gasoline prices spike, given the difficulty involved in switching sources of energy. This lack of substitutes means that consumer demand is quite unresponsive to short-term changes in the price of oil. The International Monetary Fund (IMF) has estimated that among the developed economies, represented by the Organization for Economic Cooperation and Development (OECD), [10] and non-OECD economies a 10% increase in oil prices leads to a reduction in oil demand of only 0.2%, as indicated in Table 1. Economists term such responsiveness as "elasticity," or the rate at which consumer demand changes relative to a change in prices. According to these estimates, the long-term price elasticity rises to 0.7% after 20 years for a 10% increase in oil prices. Income elasticities, however, indicate that a 10% increase in income is associated with a 6.8% increase in oil demand over the short run. Longterm, though, the income elasticity for oil drops to 2.9%, indicating that the world economy is slowly substituting away from oil. These estimates indicate that changes in oil prices, regardless of the source, have little impact on oil demand in the short run. However, the fact that the income elasticity for oil is higher in the short run than in the long run suggests that the demand for oil in the U.S. economy is more responsive to changes in national income and, therefore to policies that can affect national income. In addition, the response of oil demand to an income shock, such as during the economic recession of 2008-2009, may result in overshooting, or that demand may fluctuate more than would be expected under normal conditions, with swings in oil prices until market conditions stabilize.

Table 1. Oil Demand Price and Income Elasticities

	Short-term elasticity		Long-term elasticity	
	Price	**Income**	**Price**	**Income**
Combined OECD and non-OECD	-0.019	0.685	-0.072	0.294
OECD (developed economies)	-0.025	0.671	-0.093	0.243
Non-OECD (developing economies)	-0.007	0.711	-0.035	0.385

Source: *World Economic Outlook*, International Monetary Fund, April 2011, p. 97.

According to Global Insight,[11] a number of factors worked to put upward pressure on oil prices in 2007 and during the first half of 2008. These factors include both supply and demand issues as well as geopolitical troubles in various countries, particularly Nigeria and Iran, that created uncertainties in the market concerning the stability of oil supplies. A low rate of growth in oil supplies relative to a higher rate of growth in the demand for oil has been cited as the most important market factor behind the rise in oil prices. Saudi Arabia agreed to increase its production of oil by 300,000 barrels per day in May 2008 and by an additional 200,000 barrels per day in July 2008. Also, price movements in the oil market apparently were exaggerated somewhat by trading in the oil futures market, and other producers, especially non-OPEC producers, who had not increased their supply as had been projected. On the demand side, continued strong growth in the demand for oil in Asia and the Middle East pushed the total demand for oil to rise at a pace that has been faster than the rise in supplies. Demand in the Middle East rose at double-digit rates as a result of a boom in construction and oil consumption. In Asia, demand for oil grew rapidly in China, where the government subsidized the price of oil to consumers and the government stockpiled oil to use as substitute for coal in the Beijing area during the Olympics to reduce the level of air pollution.

Similarly, a number of factors worked to push up the prices of crude oil in early 2011. Global demand for oil remained high in 2011, especially among the non-developed economies, but high prices were expected to crimp demand in the later part of the year. On the supply side, production outages in Libya and continued political and social unrest in the Middle East increased concerns over the reliability of supplies, despite increased production from Saudi Arabia, the United Arab Emirates, and Kuwait.[12]

THE DOLLAR AND THE PRICE OF OIL

For many observers, there seems to be a direct cause and effect relationship between the depreciation in the international exchange value of the dollar and the rise in the price of oil. These observers argue that because oil is priced in dollars, a depreciation in the international exchange value for the dollar against other major currencies erodes the purchasing power of oil producers. As Figure 1 indicates, a rise in the price of oil during the January 2011 to mid-May 2011 period on a weekly basis generally has been accompanied by a depreciation in the value of the dollar relative to the Euro.

In this figure, the price of oil is represented by the price of Saudi Arabian light crude, expressed in dollars per barrel, and the value of the dollar is marked relative to the Euro, expressed in dollars per Euro, so that a rise in the value of the dollar/euro rate denotes a depreciation in the value of the dollar, since more dollars are required to purchase Euros. The two sets of data generally appear to move in the same direction, however the magnitude of the movements is not correctly represented in this type of presentation. Figure 2 shows the same data as Figure 1, although the data have been converted to index numbers so they can be represented on the same scale. As a result, it is clear that movements in the price of crude oil span a greater range than the more limited range that was experienced by the dollar/Euro exchange rate and raises questions about the relationship between movements in the price of oil and the international exchange value of the dollar. Given the limited ability of producers to increase oil supplies in the short run, large swings in the price of oil as a result of changes in market conditions (or expectations about market conditions) are not surprising.

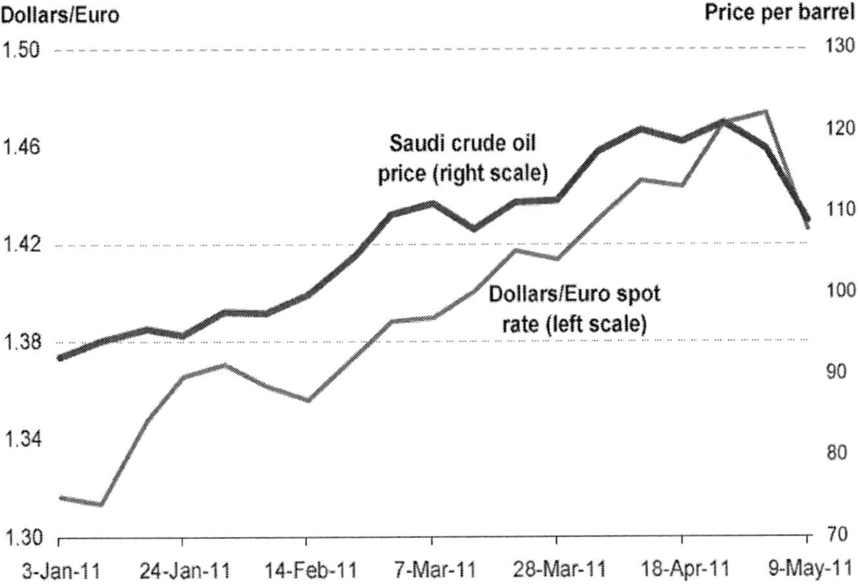

Source: Federal Reserve and the Energy Information Administration.

Figure 1. The Dollar/Euro Rate and the Price of Crude Oil, January-May 2011.

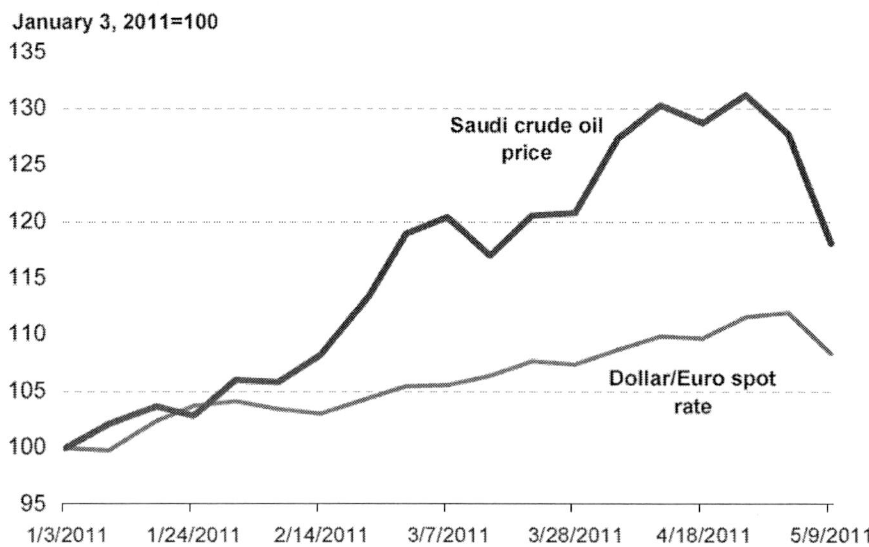

Source: Federal Reserve and the Energy Information Administration.

Figure 2. Index of the Dollar/Euro Rate and the Price of Crude Oil.

The IMF has identified three channels through which a change in the value of the dollar can affect a broad range of commodity prices, including the price of oil. A change in the value of the dollar can affect commodity prices through (1) purchasing power and cost channels; (2) asset channels in which changes in the value of the dollar affect the return on dollar-denominated financial assets; and (3) a combination of effects, including changes in monetary policy.[13] As a result of these three effects, the IMF also estimates that among various commodities, the linkage between changes in the value of the dollar and changes in commodity prices is especially strong for oil and gold, because they are more suitable as a "store of value," or as a hedge against inflation.[14] One explanation for this relationship is that oil market participants and speculators may have adopted a rough rule of thumb over time concerning changes in the value of the dollar and subsequent changes in the price of oil and vice versa. As a consequence, the statistical relationship between the two has been strengthened over time, because market participants have acted on this informal rule. In late 2010 and early 2011, however, the rise in commodity prices, including the price of oil, do not appear to be related to the value of the dollar, since commodity prices rose significantly in terms of all major currencies.

The past actions of OPEC oil producers may also have tended to strengthen the apparent linkage between changes in the value of the dollar and changes in the price of oil as the producers have acted in concert to adjust their output in order to alter the world price of oil. OPEC accounts for just over 40% of the world output of crude oil, and the coordinated actions of its members can affect world oil prices.

In addition, one of OPEC's stated goals is to secure a "fair and stable price" for the oil the member countries produce; it is not unreasonable to assume that OPEC members would respond to a loss in the purchasing power of the dollar by reducing their overall level of production, or holding down the rate of increase in production in order to raise the market price of oil.[15]

Real and Nominal Oil Prices

Figure 3 shows indexes of the nominal and real (adjusted for inflation) indexes of the price of crude oil from 1970 to 2010. The figure shows the 1973 and 1979 price increases and the slide in the real price of oil between 1980 and 1999.

The indexes show the stark rise in real oil prices in the 1970s as OPEC oil producers pushed up crude oil prices.

Over the next decade, however, real prices slowly moved downward to more moderate levels, due in part to an increase in crude oil production by non-OPEC producers.

Naturally, nominal prices increased in the 1970s as a result of the rise in oil prices, but nominal prices rose at a slower pace than real prices as national governments focused economic policies on constraining inflation. Both real and nominal oil prices began rising in 1999 as a result of an agreement signed in 1998 between OPEC members and such non-OPEC producers as Mexico, Norway, Oman, and the Russian Federation to reduce their supplies of oil. While OPEC's production of crude oil declined by about 4% in 1999 from that produced in 1998, production in 2000 increased by 6% to reach an average of 28 million barrels per day.

From 2000 to 2002, OPEC's production of crude oil fell by about 9.5% to 25.6 million barrels per day. After 2002, OPEC's crude oil production has increased every year, reaching an average of 33.9 million barrels per day in 2009.

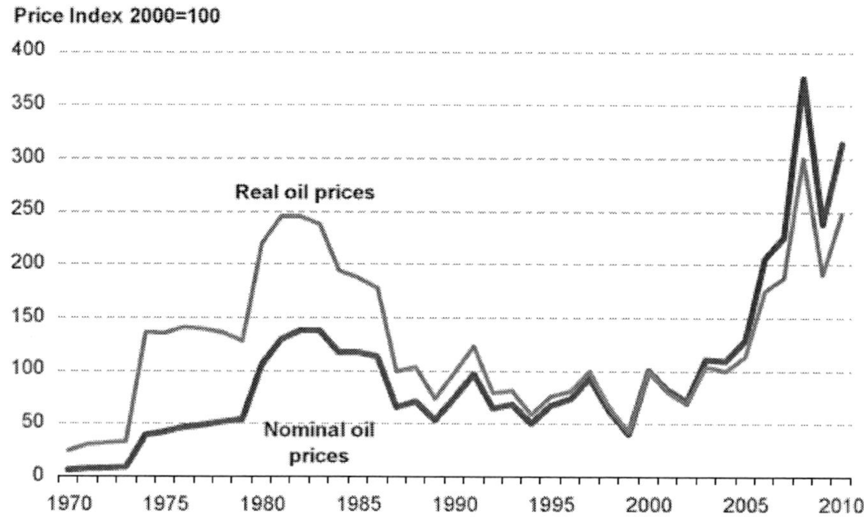

Source: CRS, Energy Information Administration.

Figure 3. Real and Nominal Crude Oil Price Indexes, 1970-2010.

Data through the fourth quarter of 2010 show that the real price of crude oil began rising during the first quarter of 2009 and rose steadily throughout the period, reaching an average price of about two-thirds of that reached during the peak period in 2008. On an annual basis, the average price of oil, as measured by the spot price of Brent crude,[16] rose from an average price of $54.42 per barrel in 2005 to an average of $96.85 per barrel in 2008, or an increase of 78% in nominal terms. During the same period, the dollar depreciated about 10% in real terms as measured against a broad basket of currencies.[17] From January 2008 to July 2008, the real price of oil increased by another 38%, while the real broad dollar index depreciated by 1.7%. Relative to other major currencies, the dollar depreciated about 7% against the Euro in real, or price adjusted, terms on average from 2005 to 2008 and about 5.6% in the January to July period in 2008. Relative to the Yen, the average value of the dollar depreciated about 15% between 2005 and 2008 in real terms, and depreciated about 5% against the Yen in the first seven months of 2008. Against the British Pound, the dollar depreciated about 8% in real terms between 2005 and 2008, and fell by about 3% in value in real terms in the first seven months of 2008. From 2009 through 2010, the dollar has appreciated slightly against a broad basket of currencies in real terms and against the pound and the euro, but depreciated against the yen in real terms.

Major Currencies

Figure 4, Figure 5, Figure 6, and Figure 7 display indexes of the dollar relative to other currencies in real terms and an index of the price of oil, also expressed in real terms, from the first quarter of 1999 through the fourth quarter of 2010. Figure 4 shows the real broad dollar index, or an index of the dollar per a unit of a grouping of 26 currencies in real terms compared with an index of the real price of crude oil. A decline in the dollar index signifies a depreciation in the value of the dollar relative to the broad group of other currencies. The data cast doubt on the argument that the price of oil responded to offset the depreciation of the dollar. Compared with the currencies of the 26 largest U.S. trading partners, the dollar has fluctuated slightly in real terms, compared with large swings in the real price of oil.

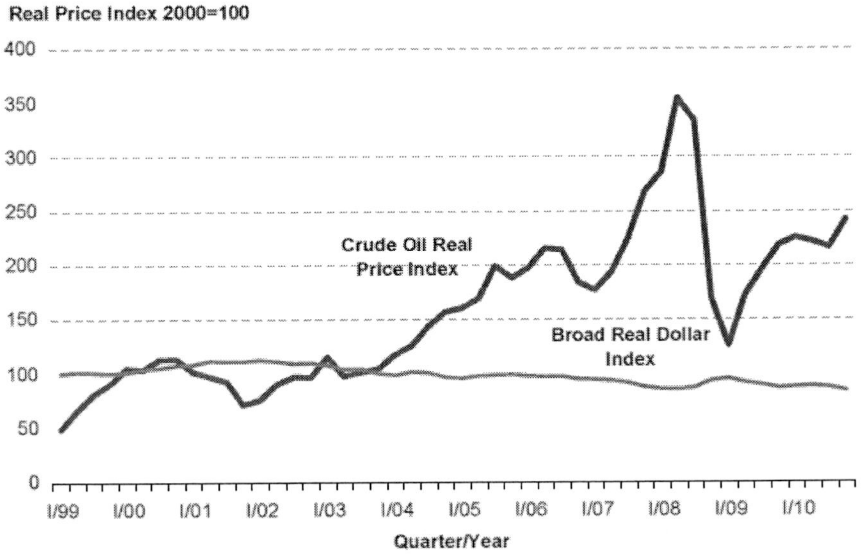

Source: Federal Reserve, Energy Information Administration.

Figure 4. Crude Oil Real Price Index and Broad Real Dollar Index,1999-2010.

The devaluation of the dollar against the Euro from early 2006 to the end of 2008 led some observers to speculate that oil producers would attempt to raise the price of oil to compensate for the devalued purchasing power of the dollar relative to the Euro and that a devalued dollar would be a disincentive for producers to explore and drill for new wells because of the loss of purchasing power. In addition, the devalued dollar makes oil cheaper for the Euro-area countries and, therefore, oil consumption in the euro area should increase with an appreciation of the euro. The decline in the exchange value of the dollar relative to the Euro also prompted some observers to argue that oil should be priced in a currency other than the dollar.

Data through fourth quarter 2010, however, do not support the contention that Euro-area countries increased their consumption of oil any faster than did the United States due to the drop in the price of oil that resulted from an appreciation of the euro relative to the dollar. Also, after some initial adjustment, pricing oil in Euros, or some other currency, rather than in dollars would appear to have no real effect on the demand and supply of oil in the market. Between the third quarter of 2009 and the third quarter of 2010, as the Euro generally depreciated relative to the dollar, oil consumption of oil in Europe rose by 2.9% compared with an increase in consumption by the United States of 4.1%. Pricing oil in dollars facilitates the smooth functioning of the oil market, because the dollar is the most widely used currency in the world for pricing, or invoicing trade, which facilitates the cross-border comparison of goods and services.[18] Figure 5 shows an index of crude oil prices in real terms and dollars per Euro in real terms, so that a rise in the dollar/Euro index signifies an appreciation in the Euro relative to the dollar, or a depreciation in the value of the dollar. The data support the argument that any loss in oil producers' purchasing power arising from a depreciation in the value of the dollar relative to the Euro was offset by a larger increase in the price of oil, which may well provide an incentive to oil producers to expand their drilling and exploration activities.

Similar trends are seen in movements in the value of the dollar relative to the Yen and the British Pound. Figure 6 shows the index of the Yen per dollar exchange rate, expressed in real terms and the index of the real price of crude oil.

In this figure, a decline in the index indicates an appreciation in the value of the Yen relative to the dollar, since fewer Yen are required to buy a dollar. Figure 7 shows the index for dollars per Pounds expressed in real terms and the index for real crude oil prices. In this case, a rise in the dollar/Pound index indicates an appreciation in the value of the Pound, since more dollars would be required to purchase a Pound.

In both cases, the relative movement in the real prices of foreign currency against the dollar has been small relative to the increase in the real price of crude oil since 2004.

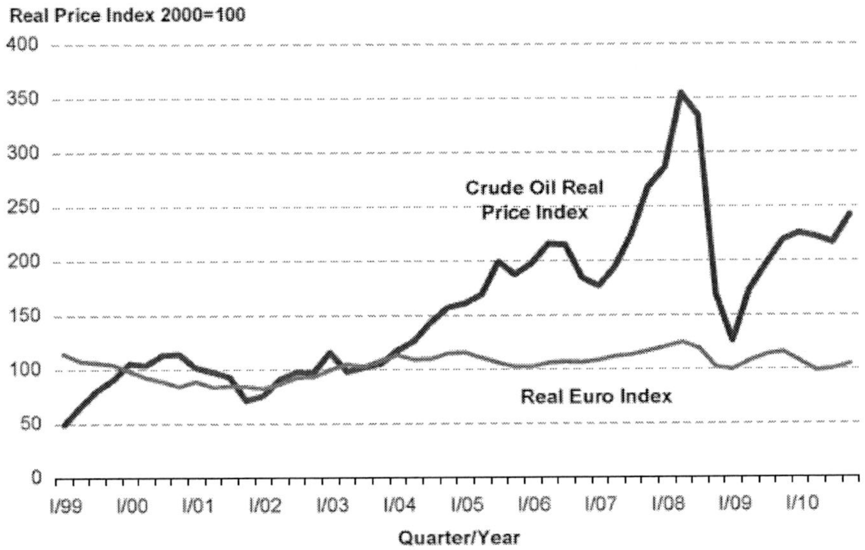

Source: Federal Reserve, Energy Information Administration.

Figure 5. Crude Oil Real Price Index and Real Dollar/Euro Index, 1999-2010.

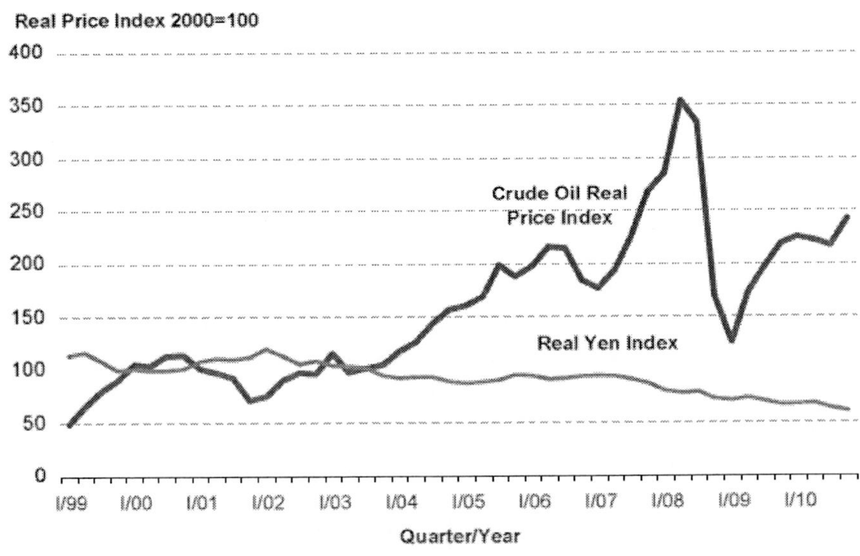

Source: Federal Reserve, Energy Information Administration.

Figure 6. Crude Oil Real Price Index and Real Yen/Dollar Index, 1999-2010.

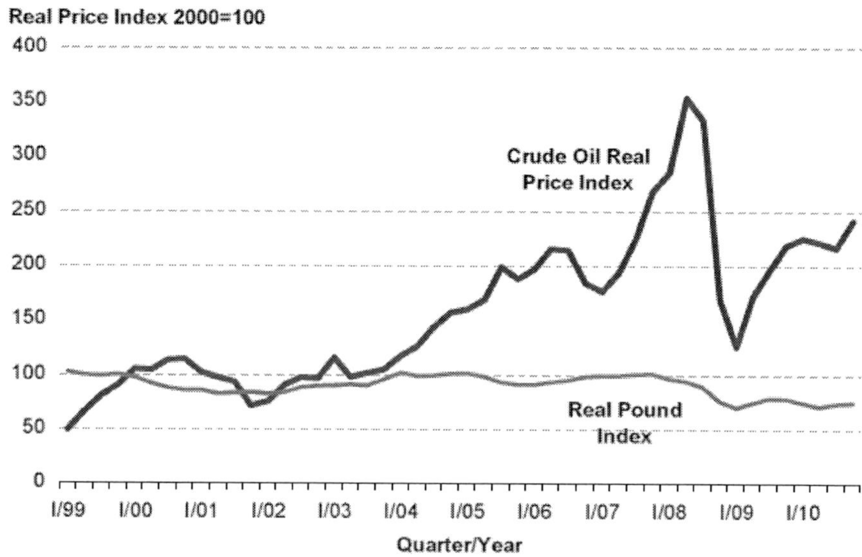

Source: Federal Reserve, Energy Information Administration.

Figure 7. Crude Oil Real Price index and Real Dollar/Pound Index, 1999-2010.

THE PRICE OF OIL

As indicated previously, the OPEC cartel of oil producers has acted in concert on occasion to alter the supply of oil in the market in order to affect the price of oil and, therefore, the export earnings of its members. In practice, OPEC oil producers, or other oil producers for that matter, do not attempt to set the price of oil directly, but attempt to alter the supply of oil in the market relative to a given level of expected demand and then rely on the market to search out the corresponding price. The price of oil, then, reflects the actual level of demand and supply in the market, which is reflected in the spot, or current, market, and the price of oil is affected by expectations about demand and supply conditions and about production capacity, reflected in the futures market. In addition, during times of economic or political instability, investors may well trade such commodities as oil that they calculate will generate a return on their investment that exceeds such traditional financial investments as stocks, bonds, or government securities.

Oil Exchanges

Similar to other commodities, oil is traded on specialized commodities exchanges. Most of this trading is conducted by licensed brokers, who act on behalf of clients to buy and sell oil on the spot market and in the futures and options markets.[19] The major futures exchanges for oil are the Intercontinental Exchange, located in London, which acquired the International Petroleum Exchange in 2001, and the New York Mercantile Exchange (NYMEX). The New York Exchange states that it is the world's largest physical commodity futures exchange. The NYMEX operates on the bid-ask system in which buy and sell transactions are executed between floor brokers. In this process, buyers compete with each other by bidding up prices and sellers compete by bidding prices down. Such markets are identified as price discovery markets, because the price of the futures contract is determined through open bids. Futures contracts are firm commitments to make or accept delivery of a specified quantity and quality of a commodity during a specified month in the future at a price agreed upon at the time the contract is made. In the commodities exchanges, futures contracts are traded in standardized units in a highly visible, extremely competitive continuous open auction. The NYMEX reports that less than 1% of all oil futures contracts take physical delivery; the remainder are settled by cash payments.

Although relatively little physical quantities of oil change hands in futures markets, the markets serve as important sources of information about market conditions and provide mechanisms for determining the price of oil in the global energy market. As a result, oil prices that are determined in the futures market are useful in at least three ways.[20] First, since the futures markets are conducted in full public view, a broad assortment of traders, including producers, commercial users, speculators, and financial institutions, make financial and production decisions based on the prices that are determined in the market. Second, the prices that are generated in the futures markets are publicly available and are used as reference points for physical trades in oil. Third, because the markets are conducted on a bid-ask system with floor brokers, the prices react quickly to new information about the supply and demand factors that are expected to influence the price of oil.

Futures and options contacts are used by both buyers and sellers to reduce the risks inherent in trading commodities.[21] Factors that might cause an abrupt change in supply, demand, and price such as international politics, war, changing economic patterns, and structural changes within the energy industry have created uncertainty about market conditions. Such uncertainty, in turn, leads to volatility in the market and creates risk for the market participants. The futures price, then, represents the current market opinion of what the commodity will be worth at some time in the future. Since the future price of a commodity can not be known with any certainty, buyers and sellers attempt to lock in prices and profit margins in advance through the use of futures and options contracts in order to hedge, or to reduce, their risks. The purpose of the hedge is to avoid the risk of an abrupt change in market conditions and prices that could result in major losses for buyers and sellers.

Since the spot price and the futures market price do not have a perfect relationship, there will always be the potential for some profit or loss. Hedging, then, reduces exposure to risk for a buyer or a seller by shifting part of the risk associated with the market price of a commodity to investors who are willing to accept the risk in exchange for a profit opportunity. As indicated above, most traders do not take physical delivery of the commodities they are trading, but hope to profit by correctly anticipating future price trends, which some observers argue has been a factor in driving high and volatile prices. Concerns over the impact of such trading on the oil market spurred a number of legislative proposals during the 110th Congress.[22]

Unlike a futures contract, an options contract conveys a right, but not an obligation, to engage in a transaction. There are two types of options, calls and puts. A call conveys the right, but not the obligation, to the one holding the option to purchase the underlying futures contract at a specified price up to a certain time. A put gives the owner of the option the right, but not the obligation, to sell the underlying futures contract at a specified price up to a certain time. A call is purchased when investors anticipate a rise in prices and a put is bought when investors expect neutral or falling prices. When options are used in combination with futures contracts, investors can develop strategies that cover virtually any risk profile, time horizon, or cost consideration.

Oil Demand and Supply

The data in Table 2 show the world demand and supply of petroleum in millions of barrels a day on average by major area from 2006 through 2010, including estimates of the four quarters of 2011. As indicated in Figure 8, between 2006 and 2010, the demand for oil, or consumption, among all consumers increased by 2.9%, rising from an average of 86.3 million barrels per day in 2006 to an average of 88.8 million barrels per day in 2010. Data for

2010 indicate that world demand for oil rose by about 1% over the 2009 amount as economic growth began recovering from the financial crisis and the global economic recession.

In 2010, the demand for oil among the developed economies increased by 0.8% and demand increased by 8.3% among the developing economies.

The developed economies, represented by the members of the OECD, accounted for about half of world demand for oil.

As a group, these developed economies decreased their demand for oil every year between 2007 and 2010. Oil demand fell in most countries and areas in 2008 as the pace of global economic activity slowed from the faster pace in 2007.

On average, oil demand among OECD countries fell by 3.6% between 2007 and 2008, compared with a 3.6% increase in demand among non-OECD countries over the same period.

During the 2006-2010 period, demand for oil by the developed OECD countries fell by 6.6% and by 7.0% in the United States. Among the developing countries, oil demand between 2006 and 2010 increased by 18.4%, led by a 30.6% increase in demand by China, although such demand started from a low base.

During the 2006-2010 period, oil supplies fell by 12.5% among the developed countries, or falling by about twice as much as the demand for oil among the developed countries, effectively increasing their demand for oil on world markets.

During the same five-year period, oil supplies among the developing countries increased by 0.9%, led by an increase in supplies by developing countries other than OPEC countries and Russia.

Figure 9 shows that world oil supplies fell by 2.8% over the period from 2005 to 2009, or by less than the increase in the world demand for oil.

During this period, oil supplies provided by U.S. producers increased by nearly 10%, while oil production in other developed countries fell by 3.9%. At the same time, OPEC producers decreased their supply of oil by 6.2%, mostly during the 2008-2009 period, and oil suppliers from other developing countries increased their supplies by 0.9%.

The shortfall between the change in demand and the change in supply was met by oil that had been held in stocks elsewhere.

The rising demand likely was an important factor in pushing up the price of oil in the market and likely affected the pricing expectations of oil brokers and traders in the futures market.

Table 2. World Oil Demand and Supply, 2006-2011 (million barrels per day)

	2006	2007	2008	2009	2010	2011 (est.)			
	Annual Average					Quarter			
						1st	2nd	3rd	4th
Petroleum (Oil) Demand									
OECD									
United States	20.69	20.68	19.50	18.77	19.23	19.25	18.97	19.47	19.23
Other OECD	15.63	28.44	28.05	26.95	26.87	27.03	26.15	26.96	27.44
Total OECD	49.34	49.59	47.87	45.72	46.10	46.28	45.12	46.43	46.67
Non-OECD									
China	7.20	7.58	7.83	8.32	9.40	9.85	10.02	9.88	10.14
Former U.S.S.R.	4.21	4.27	4.35	4.21	4.30	4.30	4.22	4.46	4.46
Other Non-OECD	23.88	24.84	25.72	26.08	28.10	28.34	29.05	29.26	28.96
Total Non-OECD	35.29	36.70	37.90	38.60	41.80	42.49	43.29	43.60	43.56
Total World Demand	84.62	86.29	85.78	84.33	87.90	88.77	88.41	90.03	90.23
Petroleum (Oil) Supply									
OECD									
United States	8.33	8.46	8.51	9.14	10.7	11.0	10.7	10.8	11.0
Other OECD	13.26	13.03	12.43	11.95	8.2	8.2	8.1	8.0	8.2
Total OECD	21.59	21.48	20.95	21.09	18.9	19.2	18.8	18.8	19.2
Non-OECD									
OPEC	35.83	34.39	35.71	33.87	34.5	35.5	NA	NA	NA
Former U.S.S.R.	12.16	12.61	12.53	12.91	13.6	13.7	NA	NA	NA
Other Non-OECD	15.02	16.07	16.32	16.52	20.30	20.30	NA	NA	NA
Total Non-OECD	63.01	63.06	64.56	63.30	68.40	69.50	NA	NA	NA
Total World Supply	84.60	84.54	85.51	84.39	87.3	88.7	NA	NA	NA
Difference (demand less supply)	0.02	1.75	0.27	-0.05	-0.60	-0.07	NA	NA	NA

Source: *International Petroleum Monthly*, December, 2010. Energy Information Administration. Table 2.1; *Oil Market Report*, International Energy Agency, April 12, 2011, Table 1.

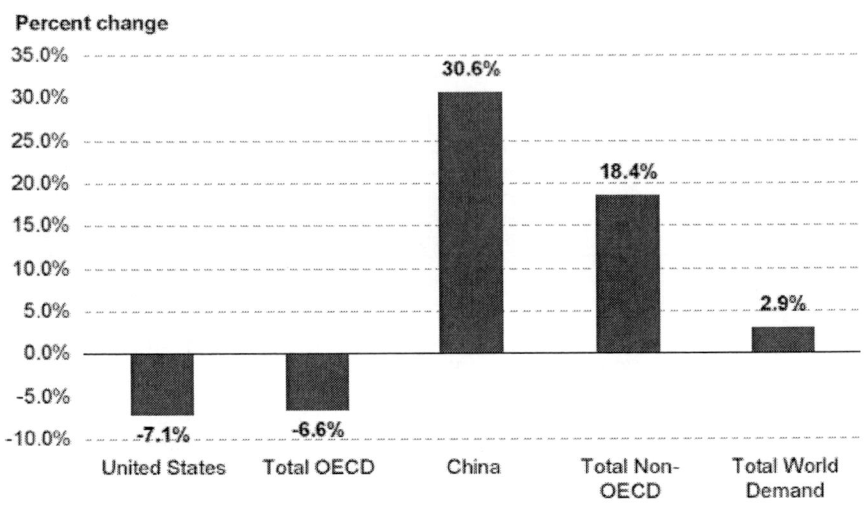

Source: Energy Information Administration.

Figure 8. Change in Oil Demand by Major Area, 2006 to 2010.

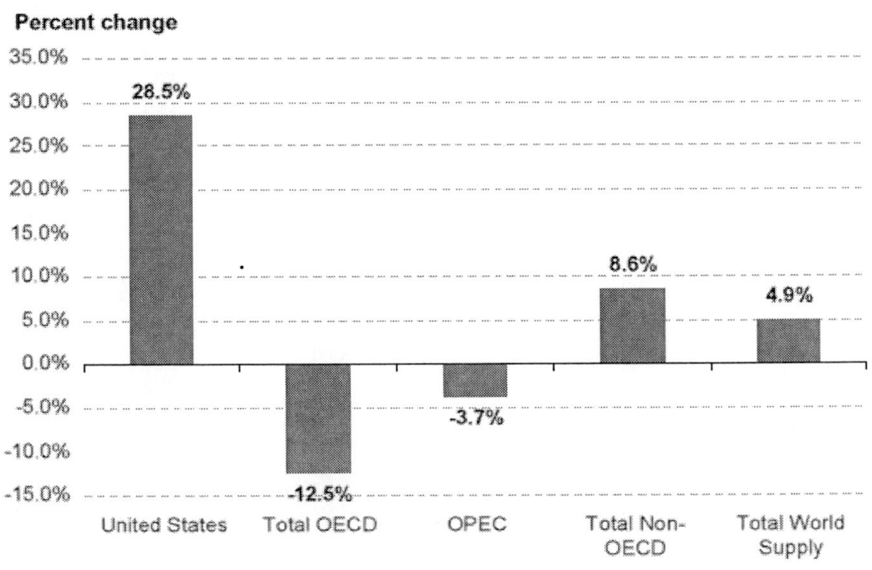

Source: Energy Information Administration.

Figure 9. Change in Oil Supply by Major Area, 2006 to 2010.

THE INTERNATIONAL EXCHANGE VALUE OF THE DOLLAR

Although attention has focused on the international exchange value of the dollar for many years, the general depreciation of the dollar since 2006 has drawn particular attention. As previously stated, some observers have argued that the rise in the price of oil has occurred in part to offset the decline in the purchasing power of oil producers as a result of the depreciation of the dollar against other major currencies. According to standard economic theory, the international exchange value of the dollar is determined by a complex interplay of demand for and supply of goods and capital within the U.S. economy and the demand for and supply of dollars in international currency markets. While dollar-related transactions generally are independent of those transactions that determine the market price of oil, there may be channels through which movements in the price of oil and changes in the value of the dollar may have spillover effects. This is especially true for the price of oil, which has a far-ranging impact on the performance of the U.S. economy and on global flows of dollars. Over time, such a connection may have become more stylized in the minds of some observers who may link changes in the price of oil to changes in the value of the dollar and vice versa. Such global capital flows, in turn, are facilitated by liberalized international capital markets and floating exchange rates, which greatly expand the amount of capital flows between countries. These flows also have sparked growth in the development and the use of financial instruments that are designed to ease the international trade of currencies and to provide investors, corporations, and financial services providers with a hedge against unpredictable changes in the value of currencies.

Capital Flows

Capital inflows also help bridge the gap in the United States between the amount of credit demanded and the domestic supply of funds. A shortfall in the domestic supply of credit relative to domestic demands for those funds tends to raise domestic interest rates and draws in capital from abroad. Those inflows, in turn, help to keep U.S. interest rates below the level they likely would reach without the inflows. The necessity to attract capital inflows, however, has complicated the conduct of economic policy. As the Federal Reserve has lowered interest rates on credit in order to stimulate economic activity and stem a slowdown in the economy, the lower interest rates have blunted capital inflows as foreign investors have sought assets in other markets where relative interest rates are higher.

Capital inflows, however, do allow the United States to spend beyond its means, including financing its trade deficit, because foreigners have been willing to lend to the United States in the form of exchanging goods, represented by U.S. imports, for such U.S. assets as stocks, bonds, and U.S. Treasury securities. Such inflows put upward pressure on the dollar, because demand for U.S. assets, such as financial securities, translates into demand for the dollar, since U.S. securities are denominated in dollars. As demand for the dollar rises or falls according to overall demand for dollar-denominated assets, the value of the dollar changes. These exchange rate changes, in turn, have secondary effects on the prices of U.S. and foreign goods, which tend to alter the U.S. trade balance. In addition, an increase in the U.S. rate of inflation tends to undermine the value of the dollar relative to other currencies, which tends to shift demand from the dollar to other currencies. At times, foreign governments have intervened in international capital markets to acquire the dollar directly or to acquire Treasury securities in order to strengthen the value of the dollar against particular currencies.

U.S. Financial Balance

The most common way of measuring capital inflows is through the U.S. balance of payments accounts. According to standard economic theory, macroeconomic developments in the U.S. economy are the major driving forces behind the magnitudes of capital flows, because the macroeconomic factors determine the overall demand for and supply of credit in the economy. Naturally, these macroeconomic conditions can be affected by changes in the price of oil, or by changes in macroeconomic policies. To the extent that changes in the price of oil alter the basic savings-investment relationship in the economy, such price changes could have long-lasting impact on the economy and on the trade balance.

One way of viewing the interaction between capital inflows and the domestic demand and supply of funds is through the domestic flow of funds accounts. These accounts measure financial flows across sectors of the economy, tracking funds as they move from those sectors that supply the capital through intermediaries to sectors that use the capital to acquire physical and financial assets.[23] Table 3 shows the major accounts in the net flow of funds in the U.S. economy from 1996 through the third quarter of 2010 The net flows show the overall financial position by sector, whether that sector is a net supplier or a net user of financial capital in the economy. Since the demand for funds in the economy as a whole must equal the supply of funds, a deficit in one sector must be offset by a surplus in another sector.

Table 3. Flow of Funds of the U.S. Economy, 1996-2010
(billions of dollars)

Year	Households	Businesses	Total	Government State and Local	Federal	ROW
1996	175.2	19.8	-196.8	-1.2	-195.6	137.9
1997	47.4	-18.3	-116.6	-47.5	-69.1	219.6
1998	128.0	-45.7	64.8	48.8	16.0	75.0
1999	-132.7	-62.6	115.3	9.9	105.4	231.7
2000	-371.0	-82.9	252.5	54.5	198.0	476.3
2001	-494.4	-82.9	233.4	35.4	198.0	485.4
2002	-304.0	8.7	-382.6	-95.6	-287.0	501.7
2003	-79.3	30.3	-546.3	-70.4	-476.4	529.4
2004	-67.9	136.8	-468.1	-33.0	-436.1	530.0
2005	-466.5	-44.8	-373.1	7.3	-380.4	712.1
2006	-512.2	-231.1	-188.9	76.1	-265.0	807.4
2007	70.5	-285.1	-345.0	-1.7	-343.3	638.5
2008	619.1	-1,003.1	-914.9	-137.3	-777.6	583.9
2009	273.8	211.9	-1,399.9	-85.1	-1,314.8	215.9
2010 I	392.5	262.0	-1,445.1	-39.5	-1,405.6	138.7
2010 II	1,091.0	12.6	-1,814.9	-48.4	-1,766.5	146.3
2010 III	252.7	181.9	-1,079.3	14.8	-1,094.1	259.4

Source: Board of Governors of the Federal Reserve System, Flow of Funds Accounts of the United States, Flows and Outstandings Third Quarter 2010, December 9, 2010.

Note: Negative values indicate a net inflow of funds, or that the demand for funds in that sector was greater than the supply of funds provided by that sector.

Generally, the household sector, or individuals, provides funds to the economy, because individuals save part of their income, while the business sector uses those funds to invest in plant and equipment that, in turn, serve as the building blocks for the production of additional goods and services. The government sector (the combination of federal, state, and local governments) can be either a net supplier of funds or a net user, depending on whether the sector is running a surplus or a deficit, respectively. The interplay within the economy between saving and investment, or the supply and uses of funds, tends to affect domestic interest rates, which move to equate the demand and supply of funds. Shifts in the interest rate also tend to attract capital from abroad, denoted by the rest of the world (ROW).

From 1999 until late in 2006, the household sector was dissaving, as individuals spent more than they earned. Part of this dissaving was offset by the government sector, which experienced a surplus from 1998 to 2001. As a result of the large household dissaving, however, the economy as a whole experienced a gap between domestic saving and investment that was filled with large capital inflows. Those inflows were particularly large in nominal terms from 2000 to 2008, as household dissaving continued and as government sector surpluses turned to historically large deficits in nominal terms. Such inflows kept interest rates below the level they would have reached without the inflows, but they put added pressure on the international exchange value of the dollar during that period.

In 2008 and 2009, capital inflows fell sharply, reflecting the global financial crisis and economic recession and the associated drop in international trade. This drop in capital inflows reflected a shift by households from dissaving to saving as concerns over the economy, with an attendant large loss in personal wealth, spurred households to pare back their consumption expenditures and to increase their personal savings. The business sector also shifted from a net supplier of funds in 2007 to a net consumer of funds as investments declined, again reflecting tight credit conditions and the drop in the rate of economic growth in the economy in 2008 and 2009. Both the federal government and state and local governments experienced a deterioration in their accounts as these sectors of the economy experienced large net deficits, reflecting the slowing rate of growth in the U.S. economy. The decrease in capital inflows combined with the slowing rate of economic growth and concerns about the stability of the financial services sector likely placed downward pressure on the exchange value of the dollar, or a devaluation of the dollar.

Foreign Exchange Market

International factors also affect the value of the dollar. The dollar is heavily traded in financial markets around the globe and, at times, plays the role of a global currency. Disruptions in this role have important implications for the United States and for the smooth functioning of the international financial system. This prominent role means that the exchange value of the dollar often acts as a mechanism for transmitting economic and political news and events across national borders, including expectations about the performance of the economy and concerns about the impact of such supply factors as the rise in the price of oil. While such a role helps facilitate a broad range of international economic and financial activities, it also means that the dollar's exchange value can vary greatly on a daily or weekly basis as it is buffeted by international events.

Table 4. Foreign Exchange Market Turnover
(Daily averages in April of the year indicated, billions of U.S. dollars)

	1995	1998	2001	2004	2007	2010
Foreign Exchange Market Turnover Instrument						
Spot transactions	494	568	386	631	1,005	1,490
Outright forwards	97	128	130	209	362	475
Foreign exchange swaps	546	734	656	954	1,714	1,765
Reporting gaps	53	61	28	107	129	NA
Total "traditional" turnover	1,190	1,527	1,239	1,934	3,324	3,981
Over the Counter Derivatives Market Turnover						
Foreign exchange instruments		97	87	140	291	NA
Interest rate instruments		265	489	1,025	1,686	2,083
Reporting gaps		13	19	55	113	NA
Total OTC turnover		375	575	1,220	1,990	2,083
Total market turnover	1,190	1,865	1,775	3,100	5,300	6,064
United States						
Foreign exchange turnover	244	351	254	461	664	817
OTC derivatives turnover		90	135	355	607	659
Total	244	441	389	816	1,271	1,506

Source: Triennial Central Bank Survey: Foreign Exchange and Derivatives Market Activity in 2010. Bank for International Settlement, September 2010.

A triennial survey of the world's leading central banks conducted by the Bank for International Settlements in April 2010 indicates that the *daily* trading of foreign currencies through traditional foreign exchange markets[24] totals about $4 trillion, up from the $3.3 trillion reported in the previous survey conducted in 2007, as indicated in Table 4. In addition to the traditional foreign exchange market, the over-the-counter (OTC)[25] foreign exchange derivatives market reported that daily turnover of interest rate and non-traditional foreign exchange derivatives contracts reached $2.1 trillion in April 2010 The combined amount of $6.1 trillion for daily foreign exchange trading in the traditional and OTC markets is more than three times the *annual* amount of U.S. exports of goods and services. The data also indicate that 85% of the global foreign exchange turnover is in U.S. dollars, slightly lower than the 85.6% share reported in a similar survey conducted in 2007.[26]

The U.S. Trade Deficit

Rising oil prices add to the Nation's trade deficit and boost the rate of change in wholesale and consumer prices, as long as the oil price increases are not offset by actions by the Federal Reserve to tighten the money supply.[27] According to data published by the Census Bureau of the Department of Commerce,[28] the prices of petroleum products over the past year have varied considerably, at times rising faster than the change in demand for those products. As a result, the price increases of imported energy-related petroleum products worsened the U.S. trade deficit in 2006,- 2008 and again in 2010. This rising cost of oil added an estimated $120 billion in 2008 and $80 billion in 2010.[29] As previously indicated, the overall U.S. demand for oil changes little with even large changes in the price of oil, so that changes in the price of imported oil are reflected in the U.S. balance of payments. As a result, increases in the price of imported oil translate into larger merchandise trade deficits, because changes in the demand for oil relative to changes in the price of oil are slight, even with large increases in the price of imported oil, and because oil imports account for a large share of U.S. imports.

Changes in oil prices also affect the cost of a broad range of goods, services, and economic activities and the changes can affect the real discretionary incomes of consumers, which has an impact on the rate of economic growth. A lower rate of economic growth, as was experienced in late 2008 and in 2009, reduces demand for oil and the price of oil falls to equate supply and demand, assuming that the supply of oil remains constant. The trade deficit also represents a transfer of wealth from the United States to the oil producers. This transfer of wealth reduces the real discretionary incomes of U.S. consumers. To the extent that the additional accumulation of wealth abroad is returned to the United States as payments for additional U.S. exports or to acquire such assets as securities or U.S. businesses, some of the negative effects could be mitigated. The data in Table 5 provide estimates of the impact different prices for imported crude oil could have on the annual U.S. trade deficit. The table also provides estimates for the increase in the trade deficit if the amount, or the volume, of imported oil declined by 3% or rose by 3% on an annual basis, as a result of changes in the demand for oil.

Table 5. Estimates of the Impact on the U.S. Trade Deficit Associated With Various Prices for Crude Oil and Changes in Oil Import Volumes

	2010		2011			
	(Actual values)		Estimated values			
	Quantity (billions of barrels)	Value (billions of dollars)	Price per barrel			
Price per barrel		$74.66	$70.00	$80.00	$90.00	$100.00
Crude oil imports	3.38	$252.18	$236.44	$270.22	$303.99	$337.77
Total energy-related Petroleum Products imports	4.28	$323.63	$299.49	$342.28	$385.06	$427.85
Change in trade deficit (in $billions)			$-24.16	$18.65	$61.43	$104.22
With 3 percent reduction in import volumes						
Crude oil imports			$229.35	$262.11	$294.87	$327.64
Total energy-related Petroleum Products imports			$290.51	$332.01	$373.51	$415.01
Change in trade deficit (in $billions)			$-33.12	$8.38	$49.88	$91.38
With 3 percent increase in import volumes						
Crude oil imports			$243.53	$278.32	$313.11	$347.90
Total energy-related Petroleum Products imports			$308.48	$352.55	$396.61	$440.68
Change in trade deficit (in $billions)			$-15.15	$28.92	$72.98	$117.05

Source: U.S. International Trade in Goods and Services February 2011, Census Bureau. Estimates developed by CRS.

According to the Census Bureau, the United States imported 4.28 billion barrels of energy-related petroleum products in 2010. Energy-related petroleum products is a term used by the Census Bureau that includes crude oil, petroleum preparations, and liquefied propane and butane gas. Crude oil comprises the largest share by far within this broad category of energy-related imports. At an average price of $74.66 per barrel, imported petroleum products cost $323 billion dollars in 2010. After subtracting U.S. exports of petroleum products, the U.S. trade deficit in petroleum products was $265 billion, or 41%

of the total trade deficit in 2010 of $646 billion. At an average price of $80 per barrel in 2011 and assuming that the amount, or the volume, of petroleum products the United States imports does not change, the U.S. trade deficit in oil in 2011 would increase by $18 billion over the deficit recorded in 2010. At an average price of $100 per barrel in 2011, the cost of imported petroleum would add $104 billion to the annual trade deficit.

Naturally, should import volumes decrease as a result of greater energy conservation or a lower rate of economic growth, the addition to the annual trade deficit would be less. If import volumes fell by 3% at a time when the average price of imported petroleum products was $100 per barrel, the addition to the annual trade deficit would be $91 billion. Should import volumes increase by 3% and oil prices rise, the deficit would increase as well.

CONCLUSION

Despite common perceptions that there is a direct cause and effect relationship between changes in the international exchange value of the dollar and the price of oil, an analysis of recent data indicates that the rise in the price of oil is being driven by an increase in demand that is exceeding the increase in supply and by political turmoil in North Africa and the Middle East. Attempts by oil producers to raise the market price of oil in order to offset the loss of purchasing power of a depreciating dollar likely would find those efforts blunted partially or in whole by the repercussions of the rise in oil prices. Increases in oil prices tend to push up prices among a broad range of goods, services, and economic activities due to the ubiquitous presence of oil as a source of energy. In addition, higher relative rates of inflation tend to undermine the exchange value of the dollar relative to other currencies, devaluing the dollar relative to other currencies and reducing the purchasing power of the dollar. Domestically, rising commodity prices reduce real incomes and lower the overall level of consumption. In turn, lower consumption reduces economic growth, which would tend to reduce the demand for oil and lead ultimately to a lower market price for oil.

The relationship between the dollar and the price of oil is complicated by the impact the price of oil can have on the rate of inflation and the rate of economic growth in the United States, the rate of economic growth and the rate of inflation in other countries, and effects on foreign currencies. For instance, rising oil prices not only raise the price of energy in the United States, but in countries around the globe. Rising prices, in turn, tend to undermine the purchasing power of national currencies. Depending on the level of domestic dependency on foreign oil, the impact of changes in oil prices can vary. Concerns over rising prices in Europe and the prospect of slowing economic growth in the Euro zone countries have tended to push down the exchange value of the Euro relative to the dollar.[30]

Upward pressure on the market price of crude oil also can come from market participants and investors who are bidding up the price of oil in an effort to invest in commodities that they calculate will generate a rate of return that exceeds that of traditional financial investments. With demand for crude oil rising faster than supplies, it is difficult for the market to determine what the future price of crude oil might be, which provides a climate that is susceptible to speculation, although there is no clear evidence that such speculation has been a major factor in the rise in crude oil prices since 2006.

Over the long run, a sustained increase in the price of energy imports could permanently alter the composition of the nation's merchandise trade deficit. Some of the impact of higher oil prices, however, could be offset if some of the dollars are returned to the U.S. economy through increased purchases of U.S. goods and services or through purchases of such other assets as securities of U.S. businesses. Some of the return in dollars likely will come through sovereign wealth funds (SWFs), or funds controlled and managed by foreign governments, as foreign exchange reserves boost the dollar holdings of such funds. Such investments likely will add to concerns about the national security implications of foreign acquisitions of U.S. firms, especially by foreign governments, and to concerns about the growing share of outstanding U.S. Treasury securities that are owned by foreigners. Over the long run it is possible for the economy to adjust to the higher prices of energy imports by improving its energy efficiency, finding alternative sources of energy, or searching out additional supplies of energy. Increased pressure is already being applied to Congress to assist in this process.

The sharp rise in prices of energy imports experienced since mid-2010 is increasing the U.S. rate of inflation and could have a slightly negative impact on the rate of economic growth in 2011. This could pose a number of policy issues for Congress. A slowdown in the rate of economic growth in the United States will lessen the demand for energy imports and could help restrain the prices of energy imports, but likely put additional pressure on the budgets at the state, local, and federal levels of government. Other important factors will be the length of political turmoil in North Africa and the Middle East, the potential impact of Atlantic hurricanes, flooding along the Mississippi river, and the impact other natural disasters could have on the production of crude oil in the Gulf of Mexico and on oil refineries. Most immediately, higher prices for energy imports will worsen the nation's merchandise trade deficit, add to inflationary pressures, and have a disproportionate impact on the energy-intensive sectors of the economy and on households on fixed incomes.

For Congress, the increase in the nation's merchandise trade deficit could add to existing inflationary pressures and complicate efforts to stimulate the economy should the rate of economic growth slow down. In particular, Congress, through its direct role in making economic policy and its oversight role over the Federal Reserve, could face the dilemma of rising inflation, which generally is treated by raising interest rates to tighten credit, and a slowing rate of economic growth, which is usually addressed by lowering interest rates to stimulate investment. A sharp rise in the trade deficit could also add to pressures for Congress to examine the causes of the deficit and to address the underlying factors that are generating that deficit. In addition, the rise in prices of energy imports could add to concerns about the nation's reliance on foreign supplies for energy imports and capital inflows and add impetus to examining the nation's energy strategy.

End Notes

[1] CRS Report RL33521, Gasoline Prices: Causes of Volatility and Congressional Response, by Carl E. Behrens and Carol Glover.

[2] Merriman, Jane, "Weak Dollar Central to Oil Price Boom," Reuters, September 26, 2007.

[3] OPEC is comprised of Algeria, Angola, Ecuador, Indonesia, Iran, Iraq, Kuwait, Libya, Nigeria, Qatar, Saudi Arabia, UAE, and Venezuela.

[4] Reed, Stanley, "How Real is OPEC's Production Cut?" BusinessWeek, September 11, 2008.

[5] Masters, Michael W., Testimony before the Committee on Homeland Security and Governmental affairs, United States Senate, May 20, 2008.

[6] Mufson, Steven, "Speculators Did Not Raise Oil Prices, Regulator Says," The Washington Post, September 12, 2008, p. D1; Staff Report on Commodity Swap Dealers & Index Traders With Commission Recommendations, Commodity Futures Trading Commission, September 2008.

[7] CRS Report RL31608, The Effects of Oil Shocks on the Economy: A Review of the Empirical Evidence, by Marc Labonte.

[8] Clifford, Catherine, Oil at 5-month Low on Shrinking Demand. CNNMoney.com., September 5, 2008; Barr, Colin, Why Cheaper Oil Signals Trouble. CNNMoney.com. September 4, 2008.

[9] Dougherty, Carter, "Fears of European Slowdown Weaken the Euro." The New York Times, August 9, 2008.

[10] For additional information about the OECD, see CRS Report RS21128, The Organization for Economic Cooperation and Development, by James K. Jackson.

[11] Market Analysis: Forecast Highlights, Global Insight, July 1, 2008.

[12] Oil Market Report, International Energy Agency, April 12, 2011.

[13] World Economic Outlook, the International Monetary Fund, April 2008. P. 46-50.

[14] The IMF estimates that a 1 percent real depreciation in the value of the dollar would result in an increase of greater than 1 percent in the price of oil over two years. Ibid., p. 50.

[15] According to standard economic theory, a reduction in the market supply of a good relative to a given level of demand will result in a higher market price for the good since the market demand would be chasing a smaller number of goods (supply), which would tend to bid up the market price of the good.

[16] Brent crude is the largest classification of crude oil. It is used to price two-thirds of internationally traded crude oil supplies.

[17] The broad dollar index is an index of the currencies of 26 largest U.S. trading partners weighted by the importance of the country as a trading partner. For additional information, see Loretan, Mico, Indexes of the Foreign Exchange Value of the Dollar, Federal Reserve Bulletin, Winter 2005. P. 1-8.

[18] Goldberg, Linda S., and Cedric Tille, The International Role of the Dollar and Trade Balance Adjustment, NBER Working Paper 12495, August 2006; and Goldberg, Linda S, and Cedric Tille, Macroeconomic Interdependence and the International Role of the Dollar, NBER Working Paper 13820, February 2008.

[19] For additional information, see CRS Report RL34555, Speculation and Energy Prices: Legislative Responses, by Mark Jickling and Lynn J. Cunningham.

[20] CRS Report RS22918, Primer on Energy Derivatives and Their Regulation, by Mark Jickling.

[21] A Guide to Energy Hedging. New York Mercantile Exchange.

[22] Ibid.

[23] Teplin, Albert M., The U.S. Flows of Funds Accounts and Their Uses, Federal Reserve Bulletin, July 2001, pp. 431-441.

[24] Traditional foreign exchange markets are organized exchanges which trade primarily in foreign exchange futures and options contracts where the terms and condition of the contracts are standardized.

[25] The over-the-counter foreign exchange derivatives market is an informal market consisting of dealers who custom-tailor agreements to meet the specific needs regarding maturity, payments intervals or other terms that allow the contracts to meet specific requirements for risk.

[26] Triennial Central Bank Survey: Foreign Exchange and Derivatives Market Activity in 2010. Bank for International Settlement, September 2010. pp. 1-2. A copy of the report is available at http://www.bis.org/publ/rpfx07.pdf

[27] Consumer Price Index: January 2011, The Bureau of Labor Statistics. P. 1.

[28] Census Bureau, Department of Commerce. Report FT900, U.S. International Trade in Goods and Services, February 11, 2011. Table 17. The report and supporting tables are available at http://www.census.gov/foreign-trade/

[29] For additional information, see CRS Report RS22204, U.S. Trade Deficit and the Impact of Changing Oil Prices, by James K. Jackson.

[30] Dougherty, Fears of European Slowdown Weaken the Euro.

In: Oil Prices   ISBN: 978-1-61942-485-2
Editors: J. C. Mullen and B. M. Lynn   © 2012 Nova Science Publishers, Inc.

Chapter 5

REAL HELP FOR AMERICAN CONSUMERS: WHO'S PROFITING AT THE PUMP?[*]

Democratic Staff
Committee on Oversight and Government Reform
United States House of Representatives
Prepared for Ranking Member
Elijah E. Cummings

EXECUTIVE SUMMARY

With gas prices now at more than \$4 per gallon, Rep. Elijah E. Cummings, the Ranking Member of the House Committee on Oversight and Government Reform, asked minority staff to examine the fundamental causes of recent price increases. Staff reviewed the work of ten different congressional committees, including the Oversight Committee, and analyzed data and information from a number of experts, including industry representatives, government officials, and academics. This report presents the results of this review.

---

[*] This is an edited, reformatted and augmented version of a Democratic Staff Committee on Oversight and Government Reform, U.S. House of Representatives publication, Prepared for Ranking Member Elijah E. Cummings, dated May 23, 2011.

The report's chief conclusion is that, in order to make the most significant impact on lowering gas prices, the Committee's primary focus should be on countering the growing impact of excessive speculation, rather than pursuing the oil industry's priorities of increasing domestic drilling or repealing safety measures put in place after the devastating BP oil spill. Experts estimate that excessive oil speculation could be inflating prices by up to 30%, while increasing domestic drilling would impact prices by only about 1%, and then only after a decade or more. Addressing excessive speculation offers the single most significant opportunity to reduce the price of gas for American consumers.

The Oil Industry Is the Most Profitable in the World

Despite the worst economic crisis since the Great Depression, oil companies have continued to make the highest profits of any industry. The top five oil companies have enjoyed profits of nearly a trillion dollars over the past ten years. They reported profits of more than $31 billion in the first quarter of FY 2011, more than 32% higher than in the first quarter of FY 2010.

Yet oil companies continue to benefit from billions of dollars in tax subsidies, as well as special deals that allow them to drill on federal lands without paying royalties. The Office of Management and Budget estimates that eliminating unnecessary tax subsidies could save more than $43 billion over the next ten years, and the Government Accountability Office reports that U.S. taxpayers may be foregoing up to $53 billion in revenues from oil companies that drill in the Gulf without paying market-rate royalties. Both industry officials and their supporters in the House have expressed support for ending this preferential treatment:

- Former Shell CEO John Hofmeister said in February, "In the face of sustained high oil prices it was not an issue—for large companies—of needing the subsidies to entice us into looking for and producing more oil."
- House Speaker John Boehner said in April, "I don't think the – the big oil companies need to have the oil depletion allowances. ... We certainly oughta take a look at it. ... And they oughta be payin' their fair share."
- House Budget Committee Chairman Paul Ryan, when asked whether he supports ending tax subsidies for oil companies, said in April, "I

agree. ... We also want to get rid of corporate welfare. And corporate welfare goes to agribusiness companies, energy companies, financial services companies. So we propose to repeal all that."

- House Oversight Committee Chairman Darrell Issa, when asked whether he supports requiring oil companies to pay fair returns on oil leases in the Gulf, said in March, "there is bipartisan support, still, to try to fix that."

Despite these statements, every legislative effort to address these problems that has come before the House in the 112th Congress has been defeated.

Countering Excessive Speculation Will Have a Direct Impact on Prices at the Pump

According to the U.S. Energy Information Administration, the price of oil has been hovering around $100 for some time. Industry officials, regulators, and outside experts have determined that these prices are artificially high in part due to the increasing role of energy speculation in the futures market. They estimate that excessive speculation may be inflating prices by up to 30%.

On May 12, Rex Tillerson, the CEO of ExxonMobil, testified before the Senate Finance Committee. When asked by Senator Maria Cantwell how much a barrel of oil would cost without excessive speculation, he responded, "Well it's pretty hard to judge but it would be, you know, when we look at it, it's going to be somewhere in the $60 to $70 range." Similarly, on April 11, Goldman Sachs warned its investment clients that speculators may be inflating the price of oil by as much as $27 a barrel.

On March 15, Bart Chilton, a Commissioner with the U.S. Commodity Futures Trading Commission (CFTC), warned that the impact of speculators is increasing rapidly. He stated: "There are now more speculative positions in commodity markets than ever before. Between June of 2008 and January 2011, futures equivalent contracts held by these types of speculators increased 64 percent in energy contracts."

The Dodd-Frank Act included provisions to counter excessive speculation, including enhanced authority for the CFTC to set position limits and raise margin limits. However, opponents of the Act, including Speaker Boehner and Chairman Issa, have sought to repeal Dodd-Frank in its entirety, cut funding to the CFTC, and delay implementation of these provisions.

Some industry experts warn that these actions could increase gas prices. According to the Executive Director of Gasoline and Automotive Service Dealers of America, "The fastest way to six dollar gasoline is to cut the funding to the CFTC."

Increasing Drilling and Repealing Safety Measures Will Not Lower Gas Prices

In contrast to addressing excessive speculation, which may be inflating prices by up to 30%, experts believe that focusing on the industry's priorities of expanding domestic drilling and repealing safety measures will have a negligible impact on gas prices.

The U.S. Energy Information Administration examined the potential impact of expanding domestic oil exploration and drilling on the outer continental shelf of the Atlantic and Pacific coasts of the United States and the Eastern and Central regions of the Gulf of Mexico. It concluded that there would be no change in prices by the year 2020, and that there would be a decrease of only 3 cents per gallon by the year 2030.

Despite this evidence, some proponents of drilling have blamed the Obama Administration for high gas prices. On April 20, 2011, for example, Chairman Issa argued that the Administration's policy toward offshore drilling "has resulted in higher prices for gas." This echoed several statements by former Alaska Governor Sarah Palin. Numerous industry and other experts have repudiated these claims. For example:

- In January, Ken Green, a resident scholar with the American Enterprise Institute, stated: "The world price is the world price. Even if we were producing 100 percent of our oil, ... [w]e probably couldn't produce enough to affect the world price of oil."
- In March, Michael Canes, the former chief economist for the American Petroleum Institute, stated, "It's not credible to blame the Obama Administration's drilling policies for today's high prices."
- Also in March, Chris Lafakis, an economist at Moody's Analytics, stated: "There is absolutely no merit to this viewpoint whatsoever."

Critics also claim that safety measures put in place after the BP oil spill are harming local economies. These claims disregard the massive economic and environmental devastation caused by oil that gushed unabated into the

Gulf for 87 days. The spill devastated the commercial and recreational fishing industries through closures—which at their peak amounted to nearly 37% of all federal waters in the Gulf—and through decreases in consumer demand for Gulf seafood.

The spill also decimated the Gulf's travel and tourism industries, which represent 46% of the Gulf's economy, generate over $100 billion annually, and are responsible for more than a million jobs.

Finally, critics have asserted falsely that the Administration has instituted a "permitorium," or a de facto moratorium, on approving drilling permits in the Gulf. This claim appears to have been created by oil industry communications officials and repeated by Members of Congress. In fact, no such "permitorium" exists. The Administration has approved 14 deepwater drilling permits, 55 shallow water permits, and two new exploration plans since the BP oil spill.

Initial delays in obtaining permits were a result of the industry's ongoing efforts to develop an adequate technology to prevent and contain exactly the type of blowout that caused the BP oil spill. On February 17, 2011, the industry announced the completion of a "subsea capping stack" to perform this function. Less than two weeks later, the Administration approved the first of multiple deepwater drilling permits issued since the BP oil spill.

I. THE OIL INDUSTRY IS THE MOST PROFITABLE IN THE WORLD

A. Oil Companies Are Making Record Profits

Despite the worst economic crisis since the Great Depression, oil companies have continued to make the highest profits of any industry in the world. On February 3, 2011, the Democratic Staff of the House Committee on Natural Resources issued a report finding that the top five oil companies have enjoyed profits of nearly a trillion dollars over the past ten years. Analyzing data from ExxonMobil, Chevron, ConocoPhillips, BP, and Shell, the report concluded that, from 2001 to 2010, these five oil companies had profits of $952 billion.[1]

Oil companies have continued to enjoy record profits during the economic recession while other industries experienced significant losses and ordered major layoffs. In the first quarter of Fiscal Year 2011, the top five oil companies reported profits of more than \$31 billion—a 32.6% increase over the first quarter of 2010.[2] Only BP—which had the most devastating oil spill in U.S. history—posted profits lower than in the same quarter last year.[3]

ExxonMobil	\$10.7 billion	69% increase
Shell	\$6.3 billion	30% increase
Chevron	\$6.2 billion	36% increase
BP	\$5.5 billion	2.1% decrease
ConocoPhillips	\$3.0 billion	44% increase
Total	**\$31.7 billion**	**32.6% increase**

Figure A. Oil Industry Profits. First Quarter 2011 Increase Over First Quarter 2010.

B. Oil Companies Receive Billions in Unnecessary Tax Breaks

Despite these record profits, oil companies continue to receive billions of dollars in tax subsidies that industry and government officials believe are unnecessary and have no effect on gas prices. In his State of the Union address on January 25, 2011, President Obama called for an end to these unnecessary subsidies. He stated:

> I'm asking Congress to eliminate the billions in taxpayer dollars we currently give to oil companies. ... I don't know if you've noticed, but they're doing just fine on their own. So instead of subsidizing yesterday's energy, let's invest in tomorrow's.[4]

This is the same position his predecessor took six years earlier. In April 2005, then-President George W. Bush stated:

> I will tell you, with \$55 oil, we don't need incentives to oil and gas companies to explore. There are plenty of incentives.[5]

Industry officials also concede that these tax subsidies are unnecessary. On November 9, 2005, the Senate Committees on Commerce, Science and Transportation and Energy and Natural Resources held a joint hearing with five oil company CEOs.

When asked by Senator Ron Wyden whether they disagreed with President Bush's statement that oil subsidies are unnecessary to encourage exploration, they all responded that they did not disagree:

Leo Raymond, CEO, ExxonMobil: No, I do not think our company has asked for any incentives for exploration.

David O'Reilly, CEO, ChevronTexaco: Agreed.

James Mulva, CEO, ConocoPhillips: In my oral comments, I said we do not need. ... Ross Pillari. CEO, BP America. He is correct.

John Hofmeister, CEO, Shell: Yes, he is.[6]

Since he left his company, former Shell CEO John Hofmeister has made clear that his position has not changed since his testimony. On February 11, 2011, he stated, "In the face of sustained high oil prices it was not an issue—for large companies—of needing the subsidies to entice us into looking for and producing more oil."[7]

On May 5, 2011, the Office of Management and Budget (OMB) estimated the savings that would be generated by eliminating several tax provisions that result in preferential treatment for oil companies.[8] OMB estimated that, from 2012 to 2021, eliminating these tax subsidies could save approximately $43.6 billion. See Figure B.

Proposed Action on Tax Provisions	Ten-Year Savings
Repeal expensing of intangible drilling costs.	$12.4 billion
Repeal deduction for tertiary injectants.	$92 million
Repeal exception to passive loss limitations for working interests in oil and natural gas properties.	$203 million
Repeal percentage depletion for oil and natural gas wells.	$11.2 billion
Repeal deduction manufacturing deduction for oil and natural gas companies.	$18.3 billion
Increase geological and geophysical amortization period for independent producers to seven years.	$1.4 billion
Total	$43.6 billion

Figure B. OMB Savings Estimate for Eliminating Oil Subsidies.

This month, the Congressional Joint Economic Committee issued a staff report calling for an end to tax subsidies for oil companies and finding that such action would not increase gas prices, as some have claimed. The report stated:

Eliminating these tax preferences, which subsidize fossil fuel production, will both reduce the federal deficit and expedite the transition to a cleaner-energy economy. Critics of repealing these subsidies argue that the targeted tax breaks spur production and lower energy prices. In reality, most of the so-called incentives have no impact on near-term production decisions, and thus repealing them would have no effect on consumer energy prices in the immediate future. Even in the longer term, the current proposed changes to these tax provisions would have little impact on global production and a negligible effect on consumer energy prices. More importantly, these subsidies failed to prevent spikes in the price of gasoline, such as the spike that occurred in 2007-08. At the same time, these tax breaks may have discouraged investment in other industries, including alternative energy sources or energy efficiency, by distorting the effective tax rate on investments in oil and natural gas.[9]

A University of Massachusetts study issued in 2009 found that investment in clean energy sectors creates roughly three times more direct and indirect jobs than when the same amount of money is spent on developing carbon-based fuels. The study reported, for example, that investing $1 million to retrofit buildings to make them more energy efficient creates three times as many jobs than a $1 million investment in oil and natural gas.[10]

C. Oil Companies Make Billions More from No-Royalty Leases

In addition to obtaining significant tax subsidies, oil and gas companies benefit from special deals that allow them to drill on federal lands without paying a fair return on royalties.

The United States enters into leases with private companies to extract oil and gas from federal lands. In return, the oil and gas companies make payments to the federal government, and the Department of the Interior manages the collection of these revenues. In February, the Government Accountability Office (GAO) released its latest "High-Risk" Report, a study issued every two years to highlight programs that are "high-risk due to their greater vulnerabilities to fraud, waste, abuse, and mismanagement."[11]

In this year's report, GAO warned that U.S. taxpayers are not receiving a "fair return" for oil and gas revenues under the terms of existing federal leases, and that correcting these deficiencies could save as much as $53 billion. Specifically, GAO faulted so-called "royalty relief" granted by Congress in the mid-1990s to encourage additional exploration at a time when oil and natural gas prices were significantly lower. Under some of these leases, oil companies pay no royalties at all. According to GAO:

> [S]pecial lower royalty rates—referred to as royalty relief—granted
> on leases issued in the deepwater areas of the Gulf of Mexico from 1996

to 2000—a period in which oil and gas prices and industry profits were much lower than they are today—could result in between $21 billion and $53 billion in lost revenue to the federal government, compared with what it would have received without these provisions.[12]

House Oversight Committee Chairman Darrell Issa has done significant work on this issue and has called for an end to these lease provisions. On October 7, 2009, he issued a staff report warning that actual shortfalls to U.S. taxpayers could be much higher. The report stated:

> Depending upon the market price of oil and natural gas, the total cost of foregone royalties could total nearly $80 billion. ... Oil and gas royalty payments represent one of the country's largest non-tax sources of revenue. Taxpayers must get every cent that is owed to them.[13]

At a hearing on March 3, 2011, Chairman Issa stated that "there is bipartisan support, still, to try to fix that."[14] To date, however, he and other Republican House Members have consistently voted against legislative efforts to correct this problem (see Section D, below).

D. Attempts to End Favorable Treatment for Oil Industry

Although House Republican leaders have repeatedly expressed support for ending tax subsidies, no-royalty leases, and other favorable treatment for the oil industry, every effort to pass legislation through the House has failed during this Congress.

On April 25, 2011, House Speaker John Boehner stated during an interview, "I don't think the – the big oil companies need to have the oil depletion allowances." He added:

> We certainly oughta take a look at it. ... We're in a time when – when the federal government is short on revenues. We need to control spending but we need to have revenues to keep the government movin'. And they oughta be payin' their fair share.[15]

Three days later, during a town hall meeting in Wisconsin on April 28, 2011, House Budget Committee Paul Ryan expressed support for ending tax subsidies for oil companies. His exchange with a constituent went as follows:

Question: The subsidy for the oil companies that the federal government gives. They've gotta stop.
Chairman Ryan: Sure.
Question: End the oil company subsidies...
Chairman Ryan: I agree.
Question: ...and you will gain a lot of that money in the red back.
Chairman Ryan: Right, so let me, one thing I didn't get into is, we're talking about reforming the safety net of the welfare system. We also want to get rid of corporate welfare. And corporate welfare goes to agribusiness companies, energy companies, financial services companies. So we propose to repeal all that.[16]

Despite these statements, every legislative effort to address these problems that has come before the House in the 112th Congress has failed. Some examples include the following:

- On February 18, 2011, House Republicans defeated, by a vote of 251 to 174, an amendment to H.R. 1 offered by Reps. Edward Markey, Maurice Hinchey, George Miller, and Lois Capps to recover up to $53 billion from federal leases in the Gulf of Mexico that allow oil companies to drill without paying royalties.[17]
- On March 1, 2011, House Republicans defeated, by a vote of 249 to 176, a Motion to Recommit offered by Rep. William Keating to H.J. Res. 44 that would have repealed taxpayer-funded subsidies for oil companies.[18]
- On April 15, 2011, House Republicans passed, by a vote of 235 to 193, H. Con. Res. 34, the Fiscal Year 2012 budget resolution sponsored by Budget Committee Chairman Paul Ryan's, which retains approximately $40 billion in tax subsidies and other favorable provisions for oil companies, while cutting investments in clean energy by approximately 70%.[19]
- On May 5, 2011, House Republicans defeated, by a vote of 241 to 171, a motion to consider Democratic legislation to prohibit the top five oil companies from receiving tax breaks for domestic manufacturing.[20]
- On May 5, 2011, House Republicans defeated, by a vote of 238 to 171, a Motion to Recommit offered by Rep. Ben Ray Luján to H.R. 1230 that would have required oil and gas produced under U.S. leases to be offered for sale only in the United States.[21]

II. COUNTERING SPECULATION WILL HAVE A DIRECT IMPACT ON PRICES AT THE PUMP

With gas prices now more than $4 per gallon, industry officials, regulators, and outside experts have determined that prices are artificially high in part due to the actions of energy speculators. They believe excessive speculation could result in prices that are inflated by as much as 30%. Addressing excessive speculation offers the single most significant opportunity to reduce the price of gas for American consumers.

A. Experts Warn Against Manipulation by Speculators

In the futures market, where energy speculation typically takes place, legally binding futures agreements are bought and sold rather than the actual commodities themselves. These agreements are called futures contracts because they provide for the delivery of particular commodities during specified future timeframes. Commodities in a futures market may be bought or sold regardless of whether they are actually used or consumed by buyers or sellers.[22]

Examples of futures exchanges include the Chicago Mercantile Exchange, the Chicago Board of Trade, the New York Mercantile Exchange, and the Intercontinental Exchange. The Commodity Futures Trading Commission (CFTC) is authorized to oversee the industry, and each of the U.S. futures exchanges operates as a self-regulatory organization, governing the conduct of its traders, brokers and the operation of the exchange.[23]

Energy futures markets generally involve two types of traders—commercial hedgers and speculators. Hedgers are airlines, refineries, railroads, or other companies that depend on commodities for their underlying business. Hedgers trade in futures to offset risk; they attempt to lock in commodity prices today for transactions in the future.[24]

In contrast, speculators seek to profit by forecasting price trends. Speculators do not typically have a consumption interest in the underlying commodities. Instead, they profit by predicting the future price of commodities. Speculators provide liquidity to the market by assuming the risks that hedgers wish to avoid. Of the two types of traders, speculators now represent the majority of participants in today's commodity futures markets.[25]

On May 12, 2011, Rex Tillerson, the CEO of ExxonMobil, testified before the Senate Finance Committee, along with the CEOs of five other major oil companies. During his testimony, he estimated that, without excessive speculation, oil would be trading at $60 to $70 a barrel instead of more than $100 a barrel. Senator Maria Cantwell asked him, "What role do you think excessive speculation in the futures market is having on elevated oil prices?" In response, Mr. Tillerson initially said "it is very difficult to precisely say what impact it has." They then had the following exchange:

> Senator Cantwell: What do you think the price would be today if it was based on fundamentals of just supply and demand?
> Mr. Tillerson: Well, again, it's, if you were to use a pure economic approach, the economists would say it would be set at the price to develop the next marginal barrel.
> Senator Cantwell: What do you think that would be today?
> Mr. Tillerson: Well it's pretty hard to judge but it would be, you know, when we look at it, it's going to be somewhere in the $60 to $70 range.[26]

On April 11, 2011, Goldman Sachs estimated that speculators may be inflating the price of oil by as much as $27 a barrel, which could result in volatility. As a result, it warned clients to pull back on energy holdings and lock in existing profits. As reported by Reuters:

> Long-term commodity bull Goldman Sachs warned clients on Monday to lock-in trading profits before oil and other markets reverse, with the bank's estimates suggesting speculators are boosting crude prices as much as $27 a barrel. ...
> Using Goldman's estimates, that indicates the total speculative premium in U.S. crude oil is currently between $21.40 and $26.75 a barrel, or about a fifth of the price.[27]

On April 18, 2011, Michael Greenberger, a former CFTC Commissioner and authority on market regulation, testified before a House Agriculture Subcommittee that speculators have artificially increased demand and unmoored prices from market fundamentals. He stated:

> [M]any company/commercial end-users, including, inter alia, Starbucks, Hershey, Lindt & Spruengli, and Delta Airlines, have now come forward demonstrating that the futures market is in complete disarray because of excessive speculation. The Commodity Markets Oversight Coalition, an independent, non-partisan and non-profit alliance

of groups that represents commodity-dependent industries, businesses and end-users, has also adopted the position that commodity prices defy market fundamentals due to excessive speculation.[28]

Writing for Forbes on May 13, 2011, Jason Raznick explained how high oil prices caused by excessive speculation may hamper the economic recovery:

> Since the price of oil barrels is currently over $100, it is clear that something beyond the laws of supply and demand is driving the high price of oil, and with it, the high price of gasoline. Well, if it isn't supply or demand that is driving the price of oil up so high, there's really only one other culprit: oil speculators. ... It is clear to any impartial observer that regulators need to clamp down in speculative trading in the oil market before the economy will ever fully recover.[29]

B. The Influence of Speculators is Increasing Rapidly

Over the past decade, and particularly in the last several years, energy speculators have been rapidly replacing traditional commodity hedgers in the commodity futures trading markets.

On March 15, 2011, CFTC Commissioner Bart Chilton delivered the keynote address to a conference on Structured Trade and Finance in the Americas. He stated:

> Economists at Oxford, Princeton, and Rice universities and many other private researchers say that speculators have had an impact on prices. ... There are now more speculative positions in commodity markets than ever before. Between June of 2008 and January 2011, futures equivalent contracts held by these types of speculators increased 64 percent in energy contracts. In June of 2008, the number of such contracts totaled 617,000. By September of 2010, they were 923,000. And, by January of this year, they had grown to 1,011,000.[30]

More recently, on April 20, 2011, Commissioner Chilton stated: "There is a Wall Street premium on gas prices today. ... Every time folks fill up their tanks, they can expect that several dollars are due to speculation."[31]

Warnings about the growing influence of oil speculators began several years ago. In 2006, the Permanent Subcommittee on Investigations of the Senate Committee on Homeland Security and Governmental Affairs issued a report entitled, "The Role of Market Speculation in Rising Oil and Gas Prices."

The report found:

> The large purchases of crude oil futures contracts by speculators have, in effect, created an additional demand for oil, driving up the price of oil to be delivered in the future in the same manner that additional demand for the immediate delivery of a physical barrel of oil drives up the price on the spot market.[32]

The Senate report also explained how the increasing role of speculators caused upwards pressure on prices. It stated:

> Although it is difficult to quantify the effect of speculation on prices, there is substantial evidence that the large amount of speculation in the current market has significantly increased prices. Several analysts have estimated that speculative purchases of oil futures have added as much as $20–$25 per barrel to the current price of crude oil, thereby pushing up the price of oil from $50 to approximately $70 per barrel.
>
> Additionally, by purchasing large numbers of futures contracts, and thereby pushing up futures prices to even higher levels than current prices, speculators have provided a financial incentive for oil companies to buy even more oil and place it in storage.[33]

In 2008, the Subcommittee on Oversight and Investigations of the House Energy and Commerce Committee held hearings entitled, "Energy Speculation: Is Greater Regulation Necessary to Stop Price Manipulation?" One hearing witness, Fadel Gheit, Managing Director and Senior Oil Analyst at Oppenheimer & Co. Inc., testified that the increasing role of speculators was causing higher oil prices. He stated:

> There were no unexpected changes in industry fundamentals in the last 12 months, when crude oil prices were below $65 per barrel. I cannot think of any reason that explains the run-up in crude oil price, beside excessive speculation. ...
>
> I do not believe the current record crude oil price is justified by market fundamentals of supply and demand. I believe the surge in crude oil price, which more than doubled in the last 12 months, was mainly due to excessive speculation and not due to an unexpected shift in market fundamentals.[34]

C. Action to Counter Excessive Speculation

On July 21, 2010, President Obama signed into law the Dodd-Frank Wall Street Reform and Consumer Protection Act. The Act included several provisions intended to address excessive speculation, and the CFTC is in the process of developing regulations to implement these provisions.

Section 737 of the Dodd-Frank Act provided the CFTC with enhanced authority to establish and set position limits for speculators.[35] This provision was designed to "ban that speculation which exceeds the need for liquidity by commercial hedges in the commodity markets."[36] Under this provision, Congress required the CFTC to implement aggregate position limits in energy commodities—including oil—within six months after enactment of the legislation. On January 26, 2011, the CFTC issued a proposed rule to begin the process of establishing position limits.[37]

Section 736 of Dodd-Frank provided the CFTC with authority to raise margin limits for speculators who invest in commodities futures.[38] Margin limits are amounts that speculators purchasing oil futures must put aside as collateral for the exchange clearinghouse.[39] Currently, margin payments are set at roughly 6% of the total value of the contract. In contrast, the Federal Reserve Board, which determines initial margin requirements for securities trades, requires a minimum margin payment of 50% of the purchase price.[40] When margin requirements were increased recently for silver futures, silver prices fell 27% in a single week.[41]

On April 21, 2011, President Obama directed Attorney General Eric Holder to create an Oil and Gas Price Fraud Working Group. The purpose of the group is to "monitor oil and gas markets for potential violations of criminal or civil laws to safeguard against unlawful consumer harm."[42] The group is a subset of the Financial Fraud Enforcement Task Force Working Group and includes representatives from the Department of Justice, the Commodity Futures Trading Commission, the Federal Trade Commission, the Department of the Treasury, and other federal agencies.

D. Support and Opposition for Solutions

On May 11, 2011, a bipartisan group of 17 Senators wrote to urge the CFTC to "take decisive action toward meeting its statutory deadline to implement new rules to protect consumers from excessive speculation and possibly manipulation in the energy futures and swaps markets." The letter stated:

> As record high volumes of financial oil speculation spike retail gasoline prices to levels unwarranted by supply and demand fundamentals, the CFTC has fallen dramatically behind in meeting the new Dodd-Frank statutory deadlines. ... [U]nder Dodd-Frank, Congress required—as opposed to merely authorized—the CFTC to implement aggregate position limits in energy commodities like oil within 180 days of the July 21, 2010 enactment date in order to "diminish, eliminate, or prevent excessive speculation." ... The intent of this provision was to utilize the lessons learned from the oil price spikes of the summer of 2008 and prevent them from happening again.[43]

Similarly, on May 16, 2011, 55 House Members sent a letter to Attorney General Eric Holder and the Chairman of the CFTC calling on both agencies to "exercise a variety of enforcement authorities to thoroughly investigate potential illegal activity in the oil futures market." With respect to the CFTC, the letter stated:

> The agency received new authority under last year's Dodd-Frank financial reform legislation to limit excessive speculation and fix market failures in petroleum markets.
>
> This authority includes the ability to establish speculative position limits and the authority to impose margin requirements to protect the financial integrity of the energy futures trading markets. The agency's efforts at implementing this authority are not proceeding fast enough— the CFTC should act swiftly to rein in dangerous activity in this market.[44]

Rep. Walter Jones has also been outspoken on this issue. In addition to being one of three Republican House Members to vote against H.R. 1, he wrote on March 9, 2011, to "strongly urge" the CFTC to implement the provisions in the Dodd-Frank Act to counter excessive oil speculation. His letter stated:

> [M]ost oil market experts agree that excessive, unnecessary speculation by Wall Street traders is part of the problem. ... Americans cannot wait until 2012; we need relief at the pump now! Further delay by the Commission leaves consumers and markets exposed to manipulation at a time when this nation can least afford it. It also threatens to derail any hope of economic recovery. On behalf of American people, I urge you to make this issue your top priority.[45]

In contrast, when the Dodd-Frank Act passed in 2010, House Republicans voted almost unanimously against it.[46] On the day it passed the Senate, then-House Minority Leader John Boehner stated: "I think it ought to be repealed."[47] After assuming control of the House in the 112th Congress, Republican leaders began taking several steps toward this goal:

- First, on January 5, 2011, Oversight Committee Chairman Darrell Issa joined seven other Members in introducing H.R. 87 to repeal the Dodd-Frank Act in its entirety.[48]
- Second, on March 9, 2011, nearly all House Republicans voted in favor of H.R. 1, which would have cut the CFTC budget by nearly one-third.[49] CFTC Chairman Gary Gensler stated that if these budget cuts were passed, "We'd have to have significant curtailment of our staff and resources. We would not be able to police ... or ensure transparent markets in futures or swaps."[50]
- Most recently, on May 18, 2011, Republicans on the House Agriculture Committee passed legislation that would delay the implementation of provisions in the Dodd-Frank Act, including regulations intended to counter excessive speculation in the oil futures market.[51]

These actions have been criticized by a variety of industry officials. For example, on March 31, 2011, Michael Fox, the Executive Director of Gasoline and Automotive Service Dealers of America, testified at a hearing before the House Committee on Natural Resources.

He stated: "The fastest way to six dollar gasoline is to cut the funding to the CFTC."[52]

III. INCREASING DRILLING AND REPEALING SAFETY MEASURES WILL NOT LOWER GAS PRICES

In contrast to addressing excessive speculation, which may be inflating gas prices by up to 30%, experts believe that efforts to expand domestic drilling or eliminate safety measures put in place after the BP oil spill would have a negligible impact on gas prices, potentially saving only pennies per gallon even after several decades.

A. ADDITIONAL DRILLING WILL HAVE MINIMAL IMPACT ON GAS PRICES

Government, industry, and outside experts have concluded that expanding domestic drilling would have a relatively insignificant effect on the price of gas for American consumers. Although expanding domestic drilling in a responsible and safe manner may offer other benefits, its effect on gas prices would be minor, according to experts. The U.S. Energy Information Administration (EIA) is the nation's foremost independent source of energy information and analyses. In 2009, EIA examined the potential impact of expanding domestic oil exploration and drilling on the outer continental shelf of the Atlantic and Pacific coasts of the United States and the Eastern and Central regions of the Gulf of Mexico. EIA issued a report concluding that there would be no changes in gas prices by the year 2020, and that there would be a decrease of only 3 cents per gallon by the year 2030.[53] The primary reason for this result is that oil is a global commodity traded on the world market. On January 4, 2011, Ken Green, a resident scholar with the American Enterprise Institute, explained that new U.S. oil production would not affect the price of crude oil domestically. He stated: "The world price is the world price. Even if we were producing 100 percent of our oil, ... [w]e probably couldn't produce enough to affect the world price of oil."[54] Philip Verleger Jr., a prominent energy economist, agreed. He stated: "Suppose the U.S. were to boost production 1 million barrels a day. ... OPEC has the capacity to cut 1 million barrels."[55] Mr. Verleger also stated that the oil industry has successfully convinced the public of the false belief that there is a connection between U.S. drilling and oil prices.[56] According to EIA, current U.S. field production of crude oil is at its highest level in almost a decade.[57] Some experts believe there is now a glut of oil on the market and that supply has overrun demand. On April 24, 2011, Rex Tillerson, CEO of ExxonMobil, stated that U.S. inventories are at "near record highs," and "there is plenty of oil on the market."[58] Despite this evidence, some proponents of drilling have blamed the Obama Administration for high gas prices. On April 20, 2011, for example, Oversight Committee Chairman Darrell Issa argued that the Administration's policy toward offshore drilling "has resulted in higher prices for gas."[59] Similarly, on May 6, 2011, former Alaska Governor Sarah Palin stated:

> But rising gas prices—there is an inherent link, David, between energy and security, energy and prosperity, and energy and freedom, and this is something that obviously our president doesn't understand because

he's doing all that he can to manipulate the U.S. supply of energy. He is diminishing and decreasing the amount of energy in our market domestically and that, of course, is resulting in prices that are rising and gas having doubled since he has been in office.[60]

Numerous industry and other experts have repudiated these claims. For example, on March 10, 2011, Michael Canes, the former chief economist for the American Petroleum Institute, stated, "It's not credible to blame the Obama Administration's drilling policies for today's high prices." He also stated:

> World oil prices are determined in a market of around 85 million barrels per day of production and consumption, while the consequences of domestic drilling, particularly in the Gulf, likely would be more in the range of several hundred thousand to one million barrels per day, and most of that production would not occur for a number of years.[61]

> On March 14, 2011, Chris Lafakis, an economist at Moody's Analytics, stated:

> There is absolutely no merit to this viewpoint whatsoever. Near-term fluctuations in gasoline prices are determined by two primary factors: crude oil prices and seasonality. Since the deepwater drilling delay applies only to exploration and production, it would take years, maybe a decade to get any amount of crude oil out of the ground and into our gas tanks.[62]

Similarly, Tom O'Donnell, a professor of Graduate International Affairs at The New School, stated:

> The amount of extra oil that the U.S. would produce, as far as affecting the world price of oil, is almost insignificant. People who say producing more oil will bring price down for Americans are missing the fact that it's a world market. For instance, oil produced in North Slope may very well go to Japan. There's not a separate market—It's a world market.[63]

B. Repealing Safety Measures Would Disregard Damage from BP Oil Spill

Despite claims from critics that safety measures put in place after the BP oil spill are harming local economies, they are designed to prevent a repeat of the massive economic and environmental devastation caused during and after the 87 days when oil gushed unabated into the Gulf of Mexico.

The explosions on the Deepwater Horizon drilling rig on April 20, 2010, killed 11 workers and resulted in the release of over 200 million gallons (4.9 million barrels) of oil into the Gulf of Mexico, creating the worst environmental disaster in the history of the United States. In January 2011, the National Commission on the BP Deepwater Horizon Oil Spill and Offshore Drilling issued a report finding that oil from the Macondo well reached more than 780 miles along the Gulf, including salt marshes, mudflats, and sand beaches.[64]

The BP oil spill devastated the fishing industry through closures—which at their peak amounted to nearly 37% of all federal waters in the Gulf—and through decreases in consumer demand for Gulf seafood.[65] According to market research commissioned by the Louisiana Seafood Promotion Board, "70% of consumers polled expressed some level of concern about seafood safety following the Gulf oil spill and 23% have reduced their consumption of seafood."[66]

Other than Alaska, the Gulf Region produces the greatest amount of seafood by volume and value in the United States.[67] In 2008, Gulf commercial fishery landings totaled 1.27 billion pounds, with dock-side values of $659 million.[68] In the same year, "the seafood industry of the Gulf States supported over 213,000 full- and part-time jobs with related income impacts of $5.5 billion."[69]

The recreational fishing industry and associated businesses, such as bait and tackle shops, restaurants, hotels, and charters, also comprise a significant portion of the economy of the region. In 2008, 5.7 million visitors and residents took 24 million fishing trips spending more than $12.5 billion on durable equipment and trips.[70]

The presence of oil in coastal estuaries and wetlands along the Gulf coast "may alter migration patterns, decrease food availability, and disrupt life cycles."[71] Many species of commercial fish and shellfish begin early stages of development in coastal areas such as estuaries and wetlands before moving offshore and becoming adults.[72] According to the National Oceanic and Atmospheric Administration (NOAA):

> Ninety-seven percent (by weight) of the commercial fish and shellfish landings from the Gulf of Mexico are species that depend on estuaries and their wetlands at some point in their life cycle. Landings

from the coastal zone in Louisiana alone make up nearly one-third (by weight) of the fish harvested in the entire continental United States.[73]

According to recent reports, the Gulf experienced a 39% decline in commercial fish landings—representing a $62 million loss in dockside sales.[74] NOAA reports that total commercial Gulf landings for shrimp species in 2010 decreased by 35.6 million pounds (27%) when compared to 2009.[75]

In addition to devastating the commercial and recreational fisheries industries, the BP oil spill decimated the Gulf's travel and tourism industries, which represent 46% of the Gulf's economy, generate over $100 billion annually, and are responsible for over a million jobs.[76] The coastal communities of Mississippi, Alabama, and northwest Florida reported that combined lodging revenue dropped 12.9% for May through August 2010 compared to 2009—a loss of $100 million for that peak summer period alone.[77] Some communities experienced more severe losses during those months with Gulf Shores and Orange Beach, Alabama, reporting a 41.5% drop in lodging revenue during those same months.[78]

C. The "Permitorium" is a Myth

Recently, critics have asserted that the Administration has instituted a "permitorium," or a de facto moratorium, on approving additional drilling permits in the Gulf. Although there is no evidence to support this claim, it appears to have been created by oil industry communications officials and repeated by Members of Congress. In fact, the Administration has approved multiple permits, and initial delays in obtaining permits were a result of the industry's ongoing efforts to develop an adequate technology to cap a blowout.

Jane Van Ryan is the Senior Communications Manager for the American Petroleum Institute. Over the past several months, she has stated on several occasions that the Obama Administration has put in place a "permitorium," or a ban on approving drilling permits in the Gulf. For example, on April 1, 2011, she stated: "The administration imposed a moratorium and a permitorium on offshore drilling."[79] In addition, on April 8, 2011, she stated:

> [The] administration has a clear record of actions, including an offshore moratorium and permitorium, that have reduced U.S. oil and natural gas development.[80]

This exact rhetoric has been repeated by Oversight Committee Chairman Darrell Issa. For example, on April 20, 2011, he stated:

> On the first anniversary of the Deepwater Horizon oil spill, the administration's de facto moratorium, or "permitorium," on offshore drilling has resulted in higher prices for gas.[81]
>
> Similarly, on May 11, 2011, he stated:
>
> As a direct result of President Obama's ban and subsequent "permitorium" against all drilling in the Gulf of Mexico, American production has dropped, and our country's energy needs have become increasingly dependent on foreign governments.[82]

Contrary to these claims, no "permitorium" exists. The Administration has approved 14 deepwater drilling permits, 55 shallow water permits, and two new exploration plans.[83] All of these were approved *after* the lifting of a six-month moratorium established to allow industry officials time to develop new technologies to prevent and contain blowouts.

Soon after the initial explosions on the Deepwater Horizon, the President ordered the Department of Interior to determine whether "additional precautions and technologies should be required to improve the safety of oil and gas exploration and production operations on the outer continental shelf."[84] On May 27, 2010, the Department issued a report known as the "Safety Measures Report" which recommended several improvements, including enhancements for "blowout preventers."[85]

Given the systemic and serious nature of problems identified in the Safety Measures Report, Interior Secretary Salazar issued a six-month moratorium on the drilling of deepwater wells in the Gulf of Mexico and Pacific Regions.[86] One of the primary reasons the Secretary ordered this suspension was that neither industry nor government possessed the technology to contain the Macondo well or respond adequately to the massive spill that resulted.[87] The absence of this technology allowed the Macondo well to flow unabated into the Gulf of Mexico for 87 days. The suspension was intended to allow time for oil companies to develop this technology.[88]

Secretary Salazar formally lifted the suspension on October 12, 2010, acknowledging that there had been significant progress in addressing drilling safety, blowout containment, and spill response.[89] Prior to resuming drilling, however, operators would have to demonstrate their ability to deploy containment resources adequate to promptly contain a blowout in deepwater.[90]

When the suspension was lifted, no operators had developed subsea containment technology to effectively contain a deepwater blowout.[91] This changed on February 17, 2011, however, when the non-profit Marine Well Containment Company announced the completion of an initial response system that included a "subsea capping stack with the ability to shut in oil flow or to flow the oil via flexible pipes and risers to surface vessels."[92] Less than two weeks later, on February 28, 2011, the Administration approved the first of multiple deepwater drilling permits issued since the BP oil spill.[93]

D. Safety Measures Are Based on Lessons Learned from BP Oil Spill

In addition to requiring oil companies to develop and deploy technologies to prevent and contain deepwater blowouts, the Administration took several other steps to enhance the safety of drilling in the Gulf. On May 22, 2010, the President announced the creation of the National Commission on the BP Deepwater Horizon Oil Spill and Offshore Drilling, an "independent, nonpartisan entity" directed to "determine the causes of the disaster, and to improve the country's ability to respond to spills, and to recommend reforms to make offshore energy production safer."[94]

After months of examination, the Commission issued a report in January 2011, concluding that the blowout of the Macondo well was preventable and occurred as a result of systemic failures by both industry and government.[95] The report stated:

> In short, the safety risks had dramatically increased with the shift to the Gulf's deepwaters, but Presidents, members of Congress, and agency leadership had become preoccupied for decades with the enormous revenues generated by such drilling rather than focused on ensuring its safety. With the benefit of hindsight, the only question had become not whether an accident would happen, but when. On April 20, 2010, that question was answered.[96]

To address these problems, the Administration began by fundamentally reorganizing the Minerals Management Service (MMS), the component of the Department of Interior charged with overseeing drilling in the Gulf. MMS had been criticized previously because it was charged with overseeing the safety of drilling, the environmental impacts caused by drilling, and the revenue generated from drilling. According to the Commission report, MMS had a "built-in incentive to promote offshore drilling in sharp tension with its mandate to ensure safe drilling and environmental protection."[97] On May 19, 2010, Secretary Salazar eliminated MMS and replaced it with the Bureau of Ocean Energy Management, Regulation and Enforcement (BOEMRE). Within BOEMRE, Secretary Salazar created three separate and independent bureaus:

- the Office of Natural Resources Revenue;
- the Bureau of Ocean Energy Management; and
- the Bureau of Safety and Environmental Enforcement.[98]

On June 21, 2010, the Secretary swore in Michael R. Bromwich, the former Inspector General for the Department of Justice, as the new Director of BOEMRE.[99] Director Bromwich ordered the agency to conduct site-specific environmental assessments for all new and revised exploration and development plans.[100] This decision was based on the Commission's finding that the practice of categorically excluding all exploration plans from review under the National Environmental Policy Act resulted in a systemic "breakdown of the environmental review process for OCS [outer continental shelf] activities."[101]

BOEMRE established guidelines, through "notices to lessees," that required operators to submit oil spill response plans that provide well-specific blowout and worst-case discharge estimates and to document the assumptions and calculations underlying those estimates.[102]

BOEMRE also issued a Drilling Safety Rule to strengthen safety standards for well control procedures, drilling equipment (including blowout preventers), and well design, including both casing and cementing procedures. In addition, the Drilling Safety Rule required independent third-party inspection and certification of various capabilities of blowout preventers, including the ability to sever drill pipes under anticipated pressures.[103]

BOEMRE also implemented a Workplace Safety Rule to require operators to put in place performance-based standards, similar to those used by regulators in the North Sea—where accident rates are much lower.[104]

Throughout this process, BOEMRE made significant efforts to conduct outreach and solicit feedback from industry stakeholders. Beginning in August 2010, BOEMRE launched a series of fact-finding forums in eight cities across the country to collect information and exchange views about deepwater drilling safety reforms, well containment, and oil spill response. BOEMRE also obtained input from 37 elected officials on the impact the oil spill had on their constituents.[105]   In   September   2010,   BOEMRE   obtained recommendations from two joint industry task forces regarding oil spill preparedness and response and subsea well control and containment, while a third industry task force made recommendations on May 17, 2010, relating to offshore drilling safety.[106] Task force participants included member companies and affiliates of the American Petroleum Institute, International Association of Drilling Contractors, Independent Petroleum Association of America, National Ocean Industries Association, and the U.S. Oil and Gas Association.[107] Many task force recommendations are reflected in the requirements issued by BOEMRE, including API standards governing the isolation of potential flow zones during well construction and safety management systems to identify, address, and manage operational safety hazards and impacts that are well-specific.[108]

Most recently, in March 2011, BOEMRE held a workshop to assist industry officials with new safety requirements associated with new exploration and drilling permits.[109] The workshop was open to all Gulf of Mexico Region Outer Continental Shelf oil and gas stakeholders and was well-attended, with approximately 200 industry participants.

As a result of input from industry and a variety of other stakeholders, Secretary Salazar announced on May 14, 2011, that the Department of Interior would be implementing a number of actions to expand drilling in a safe and responsible manner. They include the following:

1) extending drilling leases in the Gulf and certain areas off the coast of Alaska;
2) incentivizing industry to develop unused leases both on and offshore;
3) conducting lease sales in Alaska's National Petroleum Reserve;
4) fast-tracking the evaluation of oil and gas resources in the mid- and south Atlantic Ocean;
5) holding Western and Central Gulf lease sales by mid-2012; and
6) establishing a high-level, cross-agency team to create a more efficient permitting process in Alaska to ensure that health, safety, and environmental standards are met.[110]

End Notes

[1] Democratic Staff Report, House Committee on Natural Resources, *$1 Trillion in Profits and Still at the Trough: Oil and Gas in the 21st Century* (Feb. 3, 2011) (online at democrats 20Profits%20and%20Tax%20Breaks%20(2-1-11)3JD_JP_DW_FINV5.pdf)

[2] *Gas Earnings: Exxon Profits Up 69%, Shell Up 30%,* KLPW Radio (Apr. 28, 2011) (online at www.klpw.com/content/gas-earnings Chevron, *Chevron Reports First Quarter Net Income*

of $6.2 Billion, Up From $4.6 Billion in First Quarter 2010 (Apr. 29, 2010) (online at www.chevron.com/articledocuments/latest/news_204200/5968ad10-51ed-40c3-a6b6-043a 46825a82/earnings_29April2011.pdf.cvxn).

[3] *Senate Bill Squeezes Big Oil to Ease Deficit,* Reuters (May 10, 2011) (online at news.yahoo.com/s/nm/20110510/pl_nm/us_usa_bigoil_taxbreaks).

[4] The White House, *Remarks by the President in State of Union Address* (Jan. 25, 2011) (online at www.whitehouse.gov/the-press-office/2011/01/25/remarks-president-

[5] The White House, *President Addresses American Society of Newspaper Editors Convention* (Apr. 14, 2005) (online at georgewbush-whitehouse.archives.gov/news/releases/ 2005 /04/20050414-4.html).

[6] Senate Committee on Commerce, Science, and Transportation and Senate Committee on Energy and Natural Resources, *Joint Hearing on Energy Prices and Profits*, 109th Cong. (Nov. 9, 2005) (S. Hrg. 109-307) (online at frwebgate.access.

[7] *Ex-Shell CEO Says Big Oil Can Live Without Subsidies,* National Journal (Feb. 11, 2011) (online at nationaljournal.com/daily/ex-shell-ceo-says-big-oil-

[8] Office of Management and Budget, *Budget of the United States Government, Fiscal Year 2012, Terminations, Reductions and Savings*, at 52 (online at www.whitehouse.gov/ sites/default/files/omb/budget/fy2012/assets/trs.pdf) (accessed May 20, 2011).

[9] Staff Report, Senate Joint Economic Committee, *End Tax Breaks for Big Oil: Reduce the Federal Deficit Without Increasing Prices at the Pump* (May 2011) (online at jec.senate.gov/public/index.cfm?a=Files.Serve&File_id=def3390e-c933-4420-a076-19f786cd3af0) (internal footnotes omitted).

[10] University of Massachusetts, Amherst and Political Economy Research Institute, *The Economic Benefits of Investing in Clean Energy*, at 28 (June 2009) (online at www.americanprogress.org/issues

[11] Government Accountability Office, *High-Risk Series: An Update* (Feb. 2011) (GAO-11-278) (online at www.gao.gov/docsearch/featured/highrisk.html).

[12] Government Accountability Office, *High-Risk Series: An Update*, at 201 (Feb. 2011) (GAO-11-278) (online at www.gao.gov/docsearch/featured/highrisk.html).

[13] Minority Staff Report, House Committee on Oversight and Government Reform, *Teapot Dome Revisited*, at 4, 8 (Oct. 7, 2009) (online at oversight stories/Reports/2009-10-07 mmsreport.pdf).

[14] House Committee on Oversight and Government Reform, *Hearing on Waste and Abuse: The Refuse of the Federal Spending Binge* (Mar. 3, 2011).

[15] *Interview of House Speaker John Boehner,* ABC News (Apr. 25, 2011) (online at abcnews.go.com/Politics/transcript-abc-news-jonathan-karl-interviews-speaker-john/story?id=13455021).

[16] *Paul Ryan Endorses Ending Oil Subsidies, Even Though He Voted for Them,* Wonk Room (Apr. 28, 2011) (online at wonkroom.thinkprogress.org/2011/04/28/ryan-oil

[17] *Final Vote Results for Roll Call 109,* Clerk of the House (Feb. 18, 2011) (online at clerk.house.gov/evs/2011/roll109.xml).

[18] *Final Vote Results for Roll Call 153,* Clerk of the House (Mar. 1, 2011) (online at clerk.house. gov/evs/2011/roll153.xml).

[19] *Final Vote Results for Roll Call 277,* Clerk of the House (Apr. 15, 2011) (online at clerk.house .gov/evs/2011/roll277.xml).

[20] *Final Vote Results for Roll Call 293,* Clerk of the House (May 5, 2011) (online at clerk.house. gov/evs/2011/roll293.xml).

[21] *Final Vote Results for Roll Call 297,* Clerk of the House (May 5, 2011) (online at clerk.house.gov/evs/2011/roll297.xml).

[22] Congressional Research Service, *Speculation and Energy Prices: Legislative Responses* (Aug. 6, 2008) (RL34555).

[23] Congressional Research Service, *Federal Financial Services Regulatory Consolidation: An Overview* (July 10, 2008) (RL33036).

[24] Congressional Research Service, *Speculation and Energy Prices: Legislative Responses* (Aug. 6, 2008) (RL34555).

[25] *Id.*

[26] Senate Committee on Finance, *Hearing on Oil and Gas Tax Incentives and Rising Energy Prices* (May 12, 2011).

[27] *Goldman Spooks Oil Speculators with Call to Take Profit,* Reuters (Apr. 11, 2011) (online at ca.reuters.com/article/businessNews/idCATRE73A7YN20110411).

[28] House Committee on Agriculture, Subcommittee on General Farm Commodities and Risk Management, Testimony of Michael Greenberger, *Hearing on Implementing Dodd-Frank: A Review of the CFTC's Rulemaking Process* (Apr. 13, 2011).

[29] *Even Oil Companies Know That Oil Prices Are Rigged,* Forbes (May 13, 2011) (online at blogs.

[30] Commodity Futures Trading Commission, *Interconnectedness: Keynote Address of Commissioner Bart Chilton to the 13th Annual Structured Trade and Finance in the Americas Conference* (Mar. 15, 2011) (online at www.cftc.gov/PressRoom/SpeechesTestimony/ opachilton-40.html).

[31] *Obama Blames Speculators for Rising Fuel Prices,* Reuters (Apr. 20, 2011) (online at www.reuters.com/article/2011/04/20/us-usa-energy-

[32] Staff Report, Senate Committee on Homeland Security and Governmental Affairs, Permanent Subcommittee on Investigations, *The Role of Market Speculation in Rising Oil and Gas Prices: A Need to Put the Cop Back on the Beat* (June 27, 2006) (online at hsgac.senate.gov/public/_files/SenatePrint10965MarketSpecReportFINAL.pdf).

[33] *Id.*

[34] House Committee on Energy and Commerce, Subcommittee on Oversight and Investigations, Testimony of Fadel Gheit, *Energy Speculation: Is Greater Regulation Necessary to Stop Price Manipulation? – Part II,* 110th Cong. (2008) (H. Rept. 110-128).

[35] Dodd-Frank Wall Street Reform and Consumer Protection Act, Pub. L. No. 111-203 (2010).

[36] House Committee on Agriculture, Subcommittee on General Farm Commodities and Risk Management, Testimony of Michael Greenberger, *Hearing on Implementing Dodd-Frank: A Review of the CFTC's Rulemaking Process* (Apr. 13, 2011).

[37] Commodity Futures Trading Commission, *Position Limits for Derivatives,* 76 Fed. Reg. 4752 (Jan. 26, 2011).

[38] Dodd-Frank Wall Street Reform and Consumer Protection Act, Pub. L. No. 111-203 (2010).

[39] Center for American Progress, *Oil Roulette: Rising Oil Prices Harm American Families but Enrich Serial Speculators* (Apr. 28, 2011).

[40] Securities Exchange Commission, *Margin: Borrowing Money to Pay for Stocks* (Apr. 17, 2009) (online at www.sec.gov/investor/pubs/margin.htm).

[41] *Response to Volatility in Silver Takes Hold,* New York Times (May 8, 2011) (online at www.nytimes.com/2011/05/09/business/economy/09commodities.html).

[42] Department of Justice, *Attorney General Holder Announces Formation of Oil and Gas Price Fraud Working Group to Focus on Energy Markets* (Apr. 21, 2011) (online at www.justice.gov/opa/pr/2011/April/11-ag-500.html).

[43] Letter from 17 Senators to Commodity Futures Trading Commission (May 11, 2011) (online at http://wyden.senate.gov/newsroom/press/release/?id=b3c5ea66-258c-4ebc-b6ec-61821d9fbe87).

[44] Letter from 55 Members of Congress to Attorney General Eric Holder and Commodities Futures Trading Commission Chairman Gary Gensler (May 16, 2011) (online at http://braley.house.gov/images

[45] Letter from Rep. Walter B. Jones to Commodity Futures Trading Commission (Mar. 9, 2011) (online at http://jones.house.gov/News/DocumentSingle.aspx?DocumentID=228376).

[46] H.R. 4173, Dodd-Frank Wall Street Reform and Consumer Protection Act (June 30, 2010) (House vote on conference report) (online at http://thomas.gov/cgibin/bdquery/z?d111:HR04173:@@@R).

[47] *Senate Passes Wall Street Reform,* The Hill (July 15, 2010) (online at http://thehill.com/homenews/senate/109053-senate-passes-wall-st-reform

[48] H.R. 87, To Repeal the Dodd-Frank Wall Street Reform and Consumer Protection Act (introduced Jan. 5, 2011) (online at http://thomas.loc.gov/cgi-bin/bdquery/z?d112:HR00087:@@@P/|).

[49] H.R. 1, Full-Year Continuing Appropriations Act, 2011 (Mar. 9, 2011) (online at thomas.loc.gov/cgi-bin/bdquery/D?d112:1:./temp/~bdW1VJ:@@@R|/ home/LegislativeData.php|).

[50] *Regulators: Wall Street Reform at Risk,* CNN (Mar. 1, 2011) (online at http://money.cnn.com/2011/03/01/news/economy/sec_cftc_funding

[51] *House GOP Bill to Delay Financial Regulations Clears Key Hurdle,* Huffington Post (May 20, 2011) (online at www.huffingtonpost.com/2011/05/05/dodd-frank-repulican-financial-regulation_n_857917.html).

[52] House Committee on Natural Resources, *Hearing on Harnessing American Resources to Create Jobs and Address Rising Gasoline Prices* (Mar. 31, 2011) (online at http://naturalresources.house.gov/Calendar/EventSingle.aspx?EventID=228017).

[53] Energy Information Administration, *Impact of Limitations on Access to Oil and Natural Gas Resources in the Federal Outer Continental Shelf* (2009) (online at www.eia.gov/oiaf/aeo/otheranalysis/aeo_2009analysispapers/aongr.html).

[54] *As High Gas Prices Loom, New Congress Faces Pressure on Drilling,* New York Times (Jan. 4, 2011) (online at www.nytimes.com/gwire/2011/01/04/04greenwire-as-high-gasprices-loom-new-congress

[55] *Id.*

[56] *Id.*

[57] Energy Information Administration, *Domestic Oil Production Reversed Decades-long Decline in 2009 and 2010* (Apr. 27, 2011) (online at www.eia.doe.gov/oog/info/twip/twip.asp).

[58] *Oil Prices Reach Near Record Highs Despite Market Glut,* Arabnews.com (Apr. 24, 2011) (online at arabnews.com/economy/article371083.ece).

[59] Chairman Darrell E. Issa, *The Gulf Is Still Waiting for More Drilling Permits,* National Review Online (Apr. 20, 2011) (online at http://issa.house.gov/index.php?option=com _content &task=view&id=757&Itemid=92).

[60] *America's Nightly Scoreboard,* Fox News (May 6, 2011) (online at http://sarah palininformation.wordpress.com/2011/05/06/18464/).

[61] *Energy Experts Reject Claim That U.S. Drilling Policies Caused Recent Jump In Gas Prices,* Media Matters (Mar. 14, 2011) (online at mediamatters.org/research/201103140030).

[62] *Energy Experts Reject Claim That U.S. Drilling Policies Caused Recent Jump In Gas Prices,* Media Matters (Mar. 14, 2011) (online at mediamatters.org/research/201103140030).

[63] *Id.*

[64] National Commission on the BP Deepwater Horizon Oil Spill and Offshore Drilling, *Deep Water: The Gulf Oil Disaster and the Future of Offshore Drilling*, at 176 (Jan. 2011).

[65] *Id.* at 6.

[66] *Id.*

[67] Congressional Research Service, *The Deepwater Horizon Oil Spill and the Gulf of Mexico Fishing Industry* (Feb. 17, 2011).

[68] National Oceanic and Atmospheric Administration, *NOAA's Oil Spill Response: Fish Stocks in the Gulf of Mexico* (May 12, 2010).

[69] Congressional Research Service, *The Deepwater Horizon Oil Spill and the Gulf of Mexico Fishing Industry*, at 2 (Feb. 17, 2011).

[70] *Id.*

[71] *Id.* at 7.

[72] *Id.*

[73] National Oceanic and Atmospheric Administration, *Shorelines and Coastal Habitats in the Gulf of Mexico* (May 13, 2010).

[74] *The Gulf Coast One Year after the Deepwater Horizon Oil Spill,* PeterGreenberg.com (Apr. 19, 2011) (online at www.petergreenberg.com/b/The-Gulf-Coast-One-Year-After-The-Deepwater-Horizon-Oil-Spill/152605850798752191.html).

[75] Congressional Research Service, *The Deepwater Horizon Oil Spill and the Gulf of Mexico Fishing Industry*, at 5 (Feb. 17, 2011).

[76] *Oil Spill Damage Spreads through Gulf Economies,* CNNMoney.com (June 1, 2010) (online at money.cnn.com/2010/05/30/news/economy/gulf_economy/index.htm); *Update: Travel Industry Leaders Confront Gulf Crisis,* Travel Agent Central (June 21, 2010) (online at www.travelagentcentral.com/usa-southeast/update-travel-industry-

[77] *The Gulf Coast One Year after the Deepwater Horizon Oil Spill,* PeterGreenberg.com (Apr. 19, 2011) (online at www.petergreenberg.com/b/The-Gulf-Coast-One-Year-After-The-Deepwater-Horizon-Oil-Spill/152605850798752191.html).

[78] *Id.*

[79] *On Energy—Sitting Still,* Energy Tomorrow (Apr. 1, 2011) (online at blog.energy tomorrow.org/2011/04/on-energy

[80] *A Clean, Green Non Sequitor,* Energy Tomorrow (Apr. 8, 2011) (online at blog.energytomorrow.org/2011/04/a-clean-green-non-sequitur.html).

[81] *The Gulf Is Still Waiting for More Drilling Permits,* National Review (Apr. 20, 2011) (online at issa.house.gov/index.php?option=com_content&view=article&id=757%3Athe-gulf-isstill-waiting-for-more-drilling-permits&catid=23%3Aopinion-editorials&Itemid=1).

[82] Chairman Darrell E. Issa, *Issa Votes to End Obama's American Energy "Permitorium" in Favor of Lower Gas Prices and American Jobs* (May 11, 2011) (online at issa.house.gov/index.php?option=com_content&view=article&id=778:issa-votes-to-end-obamas-american-energy

[83] Bureau of Ocean Energy Management, Regulation and Enforcement, *Status of Drilling Permits & Plans Subject to Enhanced Safety and Environmental Requirements in the Gulf of Mexico* (online at www.gomr.boemre.gov/homepg/offshore/safety permits.html) (accessed May 22, 2011).

[84] Department of the Interior, *Increased Safety Measures for Energy Development on the Outer Continental Shelf, Executive Summary* (May 27, 2010).

[85] *Id.*

[86] Minerals Management Service, *Notice to Lessees and Operators of Federal Oil and Gas Leases in the Outer Continental Shelf Regions of the Gulf of Mexico and the Pacific to*

Implement the Directive to Impose a Moratorium on All Drilling of Deepwater Wells (May 30, 2010) (NTL no. 2010-N04).

[87] Secretary of the Interior, *Decision Memorandum Regarding the Suspension of Certain Offshore Permitting and Drilling Activities on the Outer Continental Shelf* (July 12, 2010).

[88] *Id.*

[89] Secretary of the Interior, *Decision Memorandum on Termination of the Suspension of Certain Offshore Permitting and Drilling Activities on the Outer Continental Shelf* (Oct. 12, 2010).

[90] Bureau of Ocean Energy Management, Regulation and Enforcement, *Statement of Compliance with Applicable Regulations and Evaluation of Information Demonstrating Adequate Spill Response and Well Containment Resources* (Nov. 8, 2010) (NTL No. 2010-N10).

[91] Bureau of Ocean Energy Management, Regulation and Enforcement, *BOEMRE Director Discusses Future of Offshore Oil and Gas Development in the U.S. at Gulf Oil Spill Series* (Apr. 19, 2011) (online at www.boemre.gov/ooc/press/2011/press0419.htm).

[92] *MWCC Announces Initial Deepwater Oil Spill Containment System,* eNewsUSA (Feb. 18, 2011) (online at enewsusa.blogspot.com/2011/02/mwcc-announces-initial-deepwater-oil.html).

[93] Bureau of Ocean Energy Management, Regulation and Enforcement, *BOEMRE Approves First Deepwater Drilling Permit to Meet Important New Safety Standards in the Gulf of Mexico* (Feb. 28, 2011) (online at www.boemre.gov/ooc/press/2011/press0228.htm).

[94] National Commission on the BP Deepwater Horizon Oil Spill and Offshore Drilling, *Deep Water: The Gulf Oil Disaster and the Future of Offshore Drilling,* at vi (Jan. 2011).

[95] *Id.* at vii.

[96] *Id.* at 84-85.

[97] *Id.* at 56.

[98] Bureau of Ocean Energy Management, Regulation and Enforcement, *Fact Sheet: The BSEE and BOEM Separation* (Jan. 19, 2011).

[99] Department of the Interior, *Salazar Swears-In Michael R. Bromwich to Lead Bureau of Ocean Energy Management, Regulation and Enforcement* (June 21, 2010).

[100] Bureau of Ocean Energy Management, Regulation and Enforcement, *BOEMRE Director Discusses Future of Offshore Oil and Gas Development in the U.S. at Gulf Oil Spill Series* (Apr. 19, 2011) (online at www.boemre.gov/ooc/press/2011/press0419.htm).

[101] Senate Committee on Energy and Natural Resources, Testimony of Oil Spill Commission Co-Chairs Graham and Reilly, *Hearing to Review the Report and Recommendations, Including Any Recommendations for Legislative Action, Issued by the National Commission on the BP Deepwater Horizon Oil Spill and Offshore Drilling,* at 12 (Jan. 26, 2011).

[102] Bureau of Ocean Energy Management, Regulation and Enforcement, *Information Requirements for Exploration Plans, Development and Production Plans, and Development Operations and Coordination Documents on the OCS* (June 18, 2010) (NTL No. 2010-N06).

[103] Bureau of Ocean Energy Management, Regulation and Enforcement, *The Drilling Safety Rule: An Interim Final Rule to Enhance Safety Measures for Energy Development on the Outer Continental Shelf* (Sep. 30, 2010) (online at www.doi.gov/news/pressreleases/loader.cfm?csModule=security/getfile&PageID=45792).

[104] Deepwater Horizon Study Group, *Preventing Accidents in Offshore Oil and Gas Operations: The U.S. Approach and Some Contrasting Features of the Norwegian Approach,* at 7 (Jan. 2011) (online at ccrm.berkeley.edu/pdfs_papers/DHSGWorkingPapersFeb16-2011/PreventingAccidents-in-OffshoreOil-and-GasOperations-MB_DHSG-Jan2011.pdf).

[105] Bureau of Ocean Energy Management, Regulation and Enforcement, *Public Forums on Offshore Drilling* (online at www.boemre.gov/forums) (accessed Mar. 28, 2011) (Sessions were held in New Orleans, Louisiana, on August 4, 2010; Mobile, Alabama, on August 10, 2010; Pensacola, Florida, on August 11, 2010; Santa Barbara, California, on August 24, 2010; Anchorage, Alaska, on August 26, 2010; Houston, Texas, on September 7, 2010; Biloxi, Mississippi, on September 10, 2010; and Lafayette, Louisiana, on September 13, 2010).

[106] Energy API, *Oil and Natural Gas Industry Calls for Offshore Safety Changes* (Sept. 7, 2010) (online at www.api.org/Newsroom/call-safety

[107] Joint Industry Oil Spill Preparedness and Response Task Force, *Draft Industry Recommendations to Improve Oil Spill Preparedness and Response* (Sept. 3, 2010).

[108] Joint Industry Task Force to Address Offshore Operating Procedures and Equipment, *White Paper: Recommendations for Improving Offshore Safety* (May 17, 2010).

[109] Bureau of Ocean Energy Management, Regulation and Enforcement, *BOEMRE to Hold Industry Workshop on Plan Processing* (Mar. 7, 2011) (online at www.boemre.gov/ooc/press/2011/press0307a.htm).

[110] Senate Committee on Energy and Natural Resources, Testimony of Secretary of the Interior Ken Salazar, *Hearing to Hear Testimony on the Following Bills Related to Oil and Gas Development: S. 516, S. 843, S. 916, S. 917* (May 17, 2011).

INDEX

D

E

F